INTRODUCTION TO ANALYTICAL CHEMISTRY FOR UNIVERSITY STUDENTS

Wilson Ngambeki William
BSc., MSc., PhD.
Senior Lecturer in Chemistry
Sebastian Kolowa Memorial University

TANZANIA EDUCATIONAL PUBLISHERS LTD

Tanzania Educational Publishers Ltd,
TEPU House,
Uganda Road, Plot No. 45, Block MDA,
Mob:0685 997583/0758 147871
Email: tepultd@yahoo.com
Website: www.tepu.co.tz
P.O. Box 1222,
Bukoba, Tanzania.

ISBN 978 9987 07 061 9

DEDICATION

I dedicate this book to my wife, Jane Barongo, our four children: Joseline, Nelson, Catherine and Cuthbert: Prof. Joseph K. Ho of University of New Mexico, USA and Dr. John Shepherd from Finland and all Chemistry Students.

TABLE OF CONTENTS

Chapter 5

Chapter 6

Chapter 7

Chapter 8
THE CHEMISTRY OF ACIDS AND BASES 141

LIST OF TABLES

LIST OF FIGURES

PREFACE

The rate at which chemical knowledge is growing at the moment is setting serious problems for lecturers / professors of undergraduate chemistry courses. The situation is specifically difficulty in Analytical Chemistry, where a couple of advances are taking place in instrumental methods of qualitative and quantitative analysis. The general goal of basic analytical chemistry is to enable a learner to identify, quantify and carry out very clear separation of the mixture of compounds. Each of these goals requires the use of differentiating techniques.

True to the concept of analytical chemistry, as the science of chemical measurement, the book begins with a development of mathematical tools which are integral parts of the art and science of chemical analysis. In this book I have carefully chosen some basic materials expected for an introductory analytical course that most curricula should have. These include analytical techniques such as homogeneous solutions, separation by electrolysis, ion exchange chromatography, crystal growth, solubility and pH, gravimetric analysis, sample preparation techniques, complex compounds formation and its analytical applications, acid-base titration, sampling, principles of chromatography, capillary electrophoresis, electro osmosis, biochemical reactivity, enzyme, separation by biochemical and complexation reaction, separation based on both mass and density, as well as capillary gel electrophoresis. Indeed, these methods have special applications in both academic and industrial laboratories, pharmaceuticals, and it is imperative for analytical chemistry students to be thoroughly acquainted with them.

It is true that elements of quantitative chemistry have been universally taught in undergraduate courses. This book intends to serve as a text that will introduce qualitative and quantitative analysis to beginners of analytical chemistry. Indeed, the main focus is on the chemical principles underlying analytical techniques rather than the techniques themselves.

The contents in this book have been intentionally kept brief because of my prejudice against voluminous texts. This will enable the student to take it to whatever place he or she will go, and thus take advantage of that opportunity to study. It is also well known that chemistry is quantitative science, and because of that, examples showing solved questions with their respective answers are given at the end of each chapter. This will allow students to spend adequate time practicing solving questions successfully in basic analytical chemistry. Furthermore, it is assumed that the students will supplement this material by a selective consultation of some of references listed at the end of each chapter.

Considering the level at which this book is intended, it is impossible to treat these topics in much detail, and the students should be aware of this limitation. Nevertheless, I would like to be very clear that I have tried to keep the book as error free as possible, but for some reasons I might not have succeeded completely. Therefore, any noted errors are my sole responsibility.

I am particularly appreciative of many helpful contributions made by Prof. Joseph K. Ho of University of New Mexico USA. I also acknowledge the help of Dr John Shepherd from Finland who tirelessly edited this book. A special word of thanks goes to my family for bearing with me during the time I was preparing this book, most importantly, my wife, Jane Barongo, to whom this work is dedicated. She provided me with constant encouragement.

I appreciate the help and support of staff of the Sebastian Kolowa Memorial University (SEKOMU) who assisted me in different ways during the preparation of this book.

Lastly I gratefully acknowledge the role of the publishers of this book, Tanzania Educational Publishers Ltd. The quality of this book is the result of their effective skills and expertise.

Chapter 1

ERROR ANALYSIS

1   Introduction
1.1   What is Analytical Chemistry?

This book intends to deal with basic principles of Analytical Chemistry. This will enable the undergraduate chemistry student to have clear knowledge on how to determine (separate and analyze) the *qualitative* and *quantitative* composition of matter.

Analytical Chemistry is a technique of recognizing different substances and determining their constituents. Knowing the latter is of great application in science because it enables a chemist to answer questions related to chemical processes. In general it deals with separation, identification, and statistical treatment of components of matter. Before we start detailed discussion on qualitative and quantitative composition of matter, the student should be equipped with basic tools for analyzing and reporting analytical data.

1.2   Mathematical Tools
1.2.1   Types of Error in Analytical Determination

It is perfectly correct to say that all measurements invariably contain error. It does not matter how much we are struggling to improve measuring techniques, we can never attain a perfect measurement. The only alternative we have is to make sure that we know what the most probable error in the measurement is and to what extent it will affect our conclusions in a given measurement. Every measurement has some uncertainty, which is called experimental error. Many times scientific conclusion can be expressed with a high or low degree of confidence, but never with complete certainty. Experimental error is classified as either *systematic* or *random*.

1.2.2   Systematic Error

What is *systematic error*, also called *determinate error*? This is repeatable if you make the measurement over again in the same way. In general, a systematic error can be discovered and corrected. For example, using pH meter that has been standardized incorrectly produces a systematic error. Suppose that the correct buffer solution used to standardize a pH meter is 7.00, while it is in fact 7.08, this will lead to a systematic error. If at all the meter is working properly, all pH reading will be 0.08 pH unit low.

When you read a pH of 5.60 the actual of the sample is 0.08 pH unit too low. When a pH of 5.60 actual read is 5.68, this means that the actual reading of the sample is 0.08 pH unit low. When you read a pH of 5.60 the actual

1

reading is 5.68. Such systematic error can be identified by using another buffer solution of known pH to test the meter. Systematic error may be positive in some cases and negative in others. The error is repeatable and, with care and cleverness, you can detect and correct it. Because the determinant errors that are often encountered are several, only the most important ones have been included hereunder. The general classes are as follows:

1.3   Instrumental Errors and Those Due to Apparatus and Reagents

(a) Balance and weights. These may include insufficient sensitivity and uncalibrated weights, etc.
(b) Volumetric apparatus. This results from uncalibrated glassware.
(c) Reagents. This may be due to the interfering impurities.
(d) Vessels and utensils. These introduce materials through attack of glassware, porecelain, etc. In addition, loss in weight of platinum crucibles when strongly heated may be a contributing factor.

1.3.1   Errors of Method

Such errors have obvious origins in the chemical or physicochemical properties of analytical systems. The errors which are physical in nature can be eliminated or reduced to very small magnitude when proper techniques are applied. Otherwise it is methodic errors, which are often inherent in the method, and no matter how carefully and skillfully the analyst works, their magnitude remains constant unless the conditions of determinations are changed. The following are the most probable sources of methodic errors; (a) Reaction failure to reach to quantitative completion (b) Dissolution of the precipitates in the solution it is precipitated and wash liquid? (c) Possibility of another substance precipitating with the reagent used (d) Decomposition or volatilization of the precipitate on heat, and (e) Side reactions. Methodic errors, being operative errors, can easily be eliminated or diminished through the adoption of a better procedure. For example, decomposition of the precipitate on heating,: if there is a reasonable range of stability, the analyst can easily control temperature to avoid decomposition. If the washing of the precipitate is properly done, this may lead to no sensible error, whereas a significant error will result if excessive amount of liquid is used. Coprecipitation errors can be reduced to negligible proportions through reprecipitating of the contaminated precipitate. There are situations where it is practically impossible to change the reaction conditions so that better conditions can be obtained, and in reality this is what is called methodic error.

1.3.2 Operative Errors

Most of these types error are physical in nature and associated with poor manipulations by an analyst during analysis. They are often independent of the instrument and utensil used, they are not contributed by the chemical properties of the reactant, and their magnitude depends more on the analyst himself or herself than on any other factor. Operative errors often take on serious proportions if the analyst is inexperienced, thoughtless and careless. Such errors are reduced to an insignificant level by being careful, skillful, and understanding the work properly. The following are typical operative errors: leaving reaction vessels uncovered and thus introducing foreign materials into solutions; loss of material from solutions through effervescence and bumping; the spilling of liquids and solids; failure to remove precipitate quantitatively from the vessels; loss in the process of transfering of the precipitates; underwashing or overwashing of precipitates; placing crucibles in the wrong position during heating so that the flame gases gain access to the interior; insufficient time of heating or the use of improper temperatures in igniting precipitate;, errors in calculations; use of non-representative sample; failure to guard against absorption of moisture by the substance weighed, etc.

1.3.3 Personal Errors

Typically, these errors originate in the constitutional inability of an individual to make certain observations accurately. A good example is for some people who lack the ability to make judgment in colour changes in titration, for example always noting volume a little just after it has passed the end point. It is important to note that such inability is a serious form of color-blindness. These errors are often quite constant in magnitude, and they are included in the person's observations.

1.4. Random Error

What is *random error*, also called *indeterminant* error? This results from limitationsion our ability to make physical measurements and from natural fluctuation in the quantity being measured. Such error often arises from small chance variations in the instrument, the system, or even from the analyst. This originates in the limited ability of the operator to control all the variables that affect the measurements. Random error has equal chance of being positive or negative. It is always present and cannot be corrected. Some examples of random error are associated with reading a scale: one person reading an instrument several times might report several different readings. Also another *indetermined error* comes from random electrical noise in an instrument. Both positive and negative fluctuations occur with approximately equal frequency and cannot be completely eliminated. Furthermore, another source of random error is due to variation in the quantity being measured. For example, if you are to measure the blood pH in your body, you would probably get different answers for blood in different parts of the body, and the pH at a given location probably would vary with time. There is always random uncertainty in the "pH of your blood" regardless or even if there is no variation in the pH meter.

1.5 Precision and Accuracy

Precision is a measure of the reproducibility of a result. The concept reproducibility is not simple as it sounds. There are different types of reproducibility. The latter is based on replicates obtained by: (1) a single observer at a given time (day and hour); (2) the same observer at another time (in spring instead of summer, in the afternoon instead of morning); (3) an observer in the same laboratory, and so on. This means reproducibility can vary to a great or lesser extent from one situation to another.

Accuracy is a measure of how close a measured value is to the 'true' value. It is important to note that the word true is in quotes because one has to measure a true value and there is error in every measurement. Measurement may be reproducible, but wrong. For example, if the sample for preparing a solution was mistakenly prepared, one will be with a wrong concentration.
It is always true to get a series of reproducible titration in the laboratory using wrong concentration and reporting an incorrect result. In this case, the precision is good, but the accuracy is poor. Conversely, it is possible to make poor reproducible measurements clustered around the correct value. In such case, the precision is poor, but the accuracy is good. It should be clearly noted that accuracy without precision is difficult, but precisiodoes not guarantee accuracy.

1.6   Absolute and Relative Uncertainty

Absolute uncertainty expresses the margin of uncertainty associated with measurements. Suppose the estimated uncertainty in reading a calibrated buret is ± 0.02 mL, one can say that ± 0.02 mL is the absolute uncertainty associated with the reading. Relative uncertainty compares absolute uncertainty with its associated measurement. The relative uncertainty of a buret reading of 10.25 ± 0.02 mL is a dimensionless quotient.

$$Relative\ uncertainty = \frac{absolute\ uncertainty}{magnitude\ of\ measuremet}$$

$$Relative\ uncertainty = \frac{0.02\ mL}{10.25\ mL}$$

$Relative\ uncertainty = 0.002$
Percentage relative uncertainty = Relative uncertainty x 100
$$= 0.002\ x\ 100$$
Percentage relative uncertainty = 0.2%

If the absolute uncertainty in reading a burette is constant at ± 0.02 mL, the percent relative uncertainty is 0.2% for a volume of 10 mL, and 0.1% for a volume of 20 mL.

Propagation of Uncertainty
Addition and subtraction
Just assume you want to make arithematic, in which experimental uncertainties, designated e_1, e_2, and e_3, are given in paratheses.

$$
\begin{aligned}
1.75\ (\pm\,0.03) &\quad - e_1 \\
+\ 1.80\ (\pm\,0.02) &\quad - e_2 \\
\underline{-\ 0.55\ (\pm\ 0.02)} &\quad - e_3 \\
\underline{3.00\ (\pm\ e_4)} &
\end{aligned}
\qquad (1.1)
$$

The arithematical answer is 3.00; the question is what is uncertainty associated with this result? For addition and substraction, the uncertainty in the answer is obtained from the absolute uncertainties of the individual terms as follows: Uncertainty in addition and substraction.

$$e_4 = \sqrt{e_1^2 + e_2^2 + e_3^2}$$

The sum in expression (1.1), we can write expression for e_4 as follows:

$$e_4 = \sqrt{(0.03)^2 + (0.02)^2 + (02)^2}$$

The absolute uncertainty e_4 is ± 0.04_1, and the answer is 3.00 ± 0.04. The point to be noted here is that uncertainty has one significant figure, though it is written initially as 0.04_1, with first insignificant figure *subscripted*. It is good to retain insignificant figures to avoid round-off error in later calculations through the number 0.04_1. The insignificant digit was subscripted just to remind us where the last significant digit should be at the conclusion of the calculations.

$$Percentage\ Relative\ Uncertainty = \frac{0.04_1 \times 100}{3.00}$$
$$= 1.4\%$$

The uncertainty, 0.04_1, is 1.4% of the result, 3.00. The subscript 4 in 1.4% is not significant. It makes sense to drop the insignificant figures now and express the final result as:

$$3.00\ (\pm\,0.04)\ (absolute\ uncertainty)$$
$$3.00\ (\pm\,1\%)\ (relative\ uncertainty)$$

Example 1.0

The volume delivered by a buret is the difference between the final and initial readings. If the uncertainty in each reading is ± 0.02 mL, what is the uncertainly in the volume delivered? **Solutions:** suppose that the initial reading is 0.05 (± 0.02) mL and the final reading is 17.80 (± 0.02) mL, what is the uncertainty in volume delivered? The actual volume derived is the difference:

$$\begin{array}{r} 17.80\ (\pm\ 0.02) \\ \underline{-0.05\ (\pm 0.02)} \\ \underline{17.75\ (\pm e)} \end{array}$$

$$e = \sqrt{(0.02)^2 + (0.02)^2}$$
$$e = \pm\ 0.03$$

Regardless of the initial and final reading, if the uncertainty in each is ± 0.02 mL, the uncertainty in volume delivered is 0.03 mL.

1.7 Significant Figures

Significant figures are the minimum number of digits needed to write a given value in scientific notation without loss of accuracy. For example, 122.9 has four significant figures and can be written as 1.229×10^2. It is often important to note that if you write 1.2290×10^2 i.e., you imply you know the value of the digit after number 9, which is not the case for the number 122.9. The number 1.2290×10^2 has five significant figures. The number 5.202×10^{-6} has four significant figures, because all four digits are necessary. The same number can be written as 0.000005202, which has four significant figures. The zeros (0) to the left of the 5 are merely holding decimal places. There are certain numbers, which seem to be ambiguous, for example 92500. The latter can mean any of the following:

9.25×10^4 3 significant figures
9.250×10^4 4 significant figures
9.2500×10^4 5 significant figures

For clear representation, one of the three numbers should be written, instead of 92500, to indicate how many figures are actually known, and that will remove unnecessary confusion in reporting such a number. Zeros are significant when they are in the following arrangement: (i) in the middle of a number or (ii) at the end of a number on the right hand side of a decimal point. The last (furthest to the right) significant figure in a measured quantity always has some associated uncertainty. The minimum uncertainty is ± 1 in the last digit.

Practice 1.0 How many significant figures are there in each number below?
(a) 1.9030 (b) 1.40×10^4 (c) 0.004050
solution (a) 5 (b) 3 (c) 4

1.8 Significant Figures in Arithmetic

If the number to be added or subtracted has an equal numbers of digits, the answer goes to the same decimal places as in any of the individual numbers.

$$1.362 \times 10^{-4}$$
$$+\ 2.222 \times 10^{-4}$$
$$3.584 \times 10^{-4}$$

If the numbers to be added do not have the same number of significant figures, we are limited by the *least certain one*: for example, in calculation of the molecular mass of KrF_2, the answer is known only to the second decimal place, because we are limited by our knowledge of the atomic mass of Kr.

$$18.998\ 4032\ (F)$$
$$+\ 18.998\ 4032\ (F)$$
$$83.80 \qquad (Kr)$$
$$121.7968064$$

The last five digits are not significant; therefore, 121.7968064 should be rounded to 121.80 as the final answer. For correct rounding up one has to look at next higher digit; in this case it is 9 and changed to 10 (i.e., 121.80) instead of rounding down, e.g.121. 7948 correctly rounded is 121.79. During addition or subtracting numbers expressed in scientific notation, we should always express all numbers with the same exponent.

$$1.632 \times 10^5 \qquad\qquad 1.632 \times 10^5$$
$$+\ 4.107 \times 10^3 \qquad\qquad +\ 0.\ 04107 \times 10^5$$
$$+\ 0.984 \times 10^6 \qquad\qquad +\ 9.84 \times 10^5$$
$$11.51 \times 10^5$$

The sum 11.51307×10^5 is rounded to 11.51×10^5 because 9.84×10^5 limits us to two decimal places.

Practice 1.1
Round the following number as indicated:
(a) 1.2357 to 4 significant figures
(b) 2.006 to 3 significant figures
Solution (a) 1.236 (b) 2.01
Example 1.1
Write each answer with correct number of digits:
(a) 2.0 + 1.4 + 3.3 + 6.8 = 13.500
(b) 106.6 – 36.3 = 70.3000
Solution (a) 13.5 (b) 70.3

1.9   Multiplication and Division

On this part we are limited by the number of digits contained in the number with the fewest significant figures:

$$3.26 \times 10^5$$
$$\underline{\times\ 1.78}$$
$$5.80 \times 10^5$$

$$4.3179 \times 10^{12}$$
$$\underline{\times\ 3.6 \times 10^{-6}}$$
$$1.6 \times 10^{-6}$$

$$34.60$$
$$\underline{\div\ 2.46287}$$
$$14.05$$

The power of 10 has no influence on the number of figures that should always be retrained.

Example 1.2

Find the molecular mass of $C_{14}H_{10}$ with correct number of significant figures.
Solution;
Multiply the atomic mass of carbon, C by 14 and atomic mass of hydrogen, H by 10 and add the product together.

$14 \times 12.017 = 168.148_8$ 6 significant figures because has 12.0107 has 6 digits
$10 \times 1.100794 = \underline{10.\ 0794}$ 6 significant figures because has 12.0107 has 6 digits
$$178.229_2$$

A reasonable answer is 178.229. The molecular mass is limited to three decimal places because the last significant digit in $14 \times 12.0107 = 168.149_8$. It is convenient to leave extra (subscripted) digits beyond the last significant figure to avoid introducing round-off error into subsequent calculation.

Example 1.3

Uncertainty in multiplication and division, in doing that, first convert all uncertainties into percentage relative uncertainties. Then calculate the error of the product or quotient as below

$$\%e_4 = \sqrt{(\%\ e_1)^2 + (\%\ e_2)^2 + (\%\ e_3)^2}$$

$$5.64 \pm e_4 = \frac{1.76\ (\pm\ 0.03) \times 1.89\ (\pm 0.02)}{0..59\ (\pm\ 3._4\ \%)}$$

First convert absolute uncertainties into percent relative uncertainties.

$$5.64 \pm e_4 = \frac{1.76\ (\pm\ 1._7\ \%) \times 1.89\ (\pm\ 1._1\ \%)}{0.59\ (\pm\ 3._4\ \%)}$$

Then find the percentage relative uncertainty of the answer by

$$\% e_4 = \sqrt{(\%1._7)^2 + (\%\ 1._1)^2 + (\%3._4)^2}$$
$$= 4._0\%$$

The answer is 5.64 ($\pm\ 4._0\%$)

To convert the percentage relative uncertainty into absolute uncertainty, find $4._0\%$ of the answer: $4._0\% \times 5.64 = 0.2_3$
The answer is $5.64 \pm 0.2_3$) finally drop the insignificant digits:
$5.6 \ (\pm 0.2)$ (absolute uncertainty)
$5.6 \ (\pm 4\%)$ (relative uncertainty)

1.10   Logarithms and Antilogarithms

The base 10 logarithm of **n** is the number **a**, whose value is such that $1 = 10^o$. The logarithm of a number is exponent to which 10 must be raised to equal that number. $n = 10^a$ means that $\textbf{log n = a}$ $\hspace{2cm}$ (1.2)

In the following case, there is no pre exponential part
$\log a + \log b = \log (ab)$
$\log a - \log b = \log (a/b)$ and

$$-\log b = \log\frac{1}{b}$$

The following general expression representing the same thing and should be remembered when one is dealing with logarithm, i.e.

$$10^{-3} = \frac{1}{10^3} = \frac{1}{1000} = 0.001$$

For example, 2 is the logarithm of 100 because $100 = 10^2$. The logarithm of 0.001 is -3, because $0.001 = 10^{-3}$ therefore, $\log 10^{-3} = -3$. In order to find the logarithm of a number with your calculator, enter the number, and press the log function. In expression 1.2, the number **n** is said to be the antilogarithm of **a**, i.e., the antilogarithm of 2 is 100 because $10^2 = 100$ and antilogarithm of -3 is 0.001 because $10^{-3} = 0.001$. Using your calculator you can find the antilogarithm of a number, enter it in your calculator and press 10^x (or antilog or INV log).

Practice 1.3
Q How would you express each answer with correct number of digits?
(a) $1.021 + 2.69 = 3.711$
(b) antilog $(-3.22) = ?$
Solution (a) 3.71 (b) 6.0×10^{-4}

1.11 Problems

1.11.1. Round each number as indicated:
$\hspace{1.5cm}$ (a) 1.2367 to 4 significant figures
$\hspace{1.5cm}$ (b) 2.052 to 2 significant figures
$\hspace{1.5cm}$ (c) 3.006 to 3 significant figures
$\hspace{1.5cm}$ (d) 2.005 to 3 significant figures
$\hspace{1.5cm}$ (e) Make a distinction between operative and personal errors.

1.11.2. Indicate how many significant figures there are in the following:
    (a) 0.00406 (b) 0.2050 (c) 1.03×10^5

1.11.3. Write each answer with the correct number of digits:
    (a) $2.0 + 1.1 + 3.4 + 8.2 = 14.7000$
    (b) $98.9 - 24.4 = 74.500$
    (c) $(26.14/37.62) \times 2.19 = 1.52$
    (d) $\log (3.98 \times 10^4) = 4.5999$

1.11.4. (a) With examples, explain the difference between random and systematic errors. State whether error in (b)-(d) are random or systematic:

(b) A 10 mL burette consistently delivers 1.98 ± 0.01 mL when drained from exactly 0 to exactly 2 mL and consistently delivers 2.05 ± 0.02 mL when drained from 2 to 4 mL.

(c) A 50 mL transfer pipette consistently delivers 50.031 ± 0.009 mL when drained from the mark.

(d) A 25 mL burette consistently delivered 2.9839 g of water when drained from exactly 0.00 to 3.00 mL. The next time I delivered water from 0.00 to 3.00 mL mark, the delivered mass was 2.990 g.

(e) A clean funnel that had been in a drawer in the laboratory since last year had a mass 12.4329 g. When filled with solid precipitate and dried thoroughly in the oven at 110 ºC, the mass was 12. 8456 g. The mass of precipitate can be calculated as $12.8456 - 12.4329 = 0.4127$ g. Is there systematic or random error (or both) in the mass of precipitate?

1.11.5. (a) Why do we use quotation marks around the word true in the statement that accuracy refers to how close a measurement is to the "true" value?

(b) What are most probable sources of methodic errors?

1.11.6. Write each answer with the correct number of digits. Find the absolute uncertainty and percent relative uncertainty for each answer.
    (a) $6.2 (\pm 0.2) - 4.1 (\pm 0.1) = ?$
    (b) $9.43 (\pm 0.05) \times 0.016 (\pm 0.001) = ?$

1.11.7. Suppose that in a gravimetric analysis that unfortunately you have forgotten to dry the filter crucible before collecting precipitate. After filtering the product, you dry the product and crucible thoroughly before weighing them. Is the error in the mass that you have reported a systematic or random error? Is the mass of product often high or low?

1.11.8. To prepare a solution of NaCl, one student weighed $2.634 (\pm 0.002)$ g and dissolved in volumetric flask whose volume is $100.00 (\pm 0.08)$ mL. Express the molarity of the solution, along with its uncertainty, with an appropriate number of digits.

1.11.9. Using a procedure called standardization we can measure the concentration of HCl solution by reaction with pure sodium carbonate:

A volume of 27.35 ± 0.0.04 mL of HCl solution was required for complete reaction with 0.9674 ± 0.0009 g of Na_2CO_3 (FM 105.988 ± 0.001). Find the molarity of HCl and its absolute uncertainty.

1.11.10. What is uncertainty in the molecular weight of O_2? Given that the atomic mass of oxygen is 15.999 ± 0.0003 g/mol.

1.11.11. Express the absolute uncertainty in the following

$$\frac{3.43\ (\pm\ 0.08)\ x\ 10^{-8}}{2.11\ (\pm\ 0.04)\ x\ 10^{-3}}$$

1.11.12. To help identify an unknown mineral sample given to you in chemistry lecture, you are given its mass and volume, which are 4.635 ± 0.002 g and 1.13 ± 0.05 mL respectively.

  (a) Calculate percentage uncertainty in both mass and volume.
  (b) Write the value of density and its uncertainty with the correct number of digits.

1.12 References

Bilo, E. J. (2001). *Microsoft Excel for Chemists*, 2nd ed. New York: Wiley.

de Levie, R. (2001). *How to use Excel in Analytical Chemistry and in General Scientific Data Analysis,* Cambridge University Press.

Dodge, J. R. (2003). The Exford Dictionary of Statistical Terma, OUT.

Edmiston, P. L. and Williams, T. R. (2000). An Analytical Laboratory Experiment in error Analysis: Repeated Determination of Glucose Using Commercial Glucometers, 77, 377.

Harris, D. C. (2003). *Quantitative Chemical Analysis*, 6rd ed. W. H. Freeman and Company, New York NJ,

Harris, D. C. (2005). *Exploring Chemical Analysis*, 3rd ed. W. H. Freeman and Company, New York NJ.

Paakkunainen, M., Kilpelainen, Y. Reinkainen, S. and Minkkinen, P. (2007). Effect of Missing Valuesin Estimation of Meanof Auto-correlated Measurement Series, *Analytical Chimica Acta*, 595, 209.

Tylot, J. R. (1999). An Introduction to Error Analysis: The Study of Uncertainities in Physical Measurements University Science Book.

Chapter 2

QUALITATIVE ANALYSIS

2.1   Introduction

Qualitative chemical analysis deals essentially with the detection and identification of the constituents of a given sample. The word sample means that portion of material, which is to be examined. The constituent to be determined or detected may be: (i) elements in or without regard to their state combination or oxidation, (ii) radicals, (iii) functional groups (as in organic substances such as OH^-, $-COR$, $COOH^-$), (iv) compounds or even phases, which may or may not be compounds. In some cases the qualitative composition of a sample may be unknown or imperfectly known and should always be analyzed qualitatively before considering the aspect of quantitative analysis.

Qualitative tests may entail identification of the form of an element, for example whether copper is present as Cu(II) or Cu(I); nitrate or nitrite, Fe(II) or Fe(III) oxide is present. Generally speaking, qualitative tests are to some extent quantitative and provide an indication of the amount of reacting constituent, whether by the quantity of a precipitate, intensity of a colour, etc. A *qualitative analysis* provides a rough indication of the amounts of the constituents present and because of that this step is of great value in making the choice of appropriate methods in determining those constituents. On the other hand, qualitative analysis may provide a rough indication of the amounts of unexpected elements in a sample, which could lead to gross errors if disregarded in the quantitative procedure.

Under qualitative analysis different techniques are used to identify a variety of substances, for example:

Atomic Spectroscopy - it is used to identify the amount of metallic elements particularly metals at major and trace levels in industrial research and environmental laboratories.

Infrared Spectroscopy - this is used to identify organic functional groups e.g., reactive atoms or groups, for example:

12

Microscopic examination (especially with a polarizing microscope) - shows whether a finely divided solid is homogeneous or heterogeneous, and *X - ray Diffraction* - is used for identification of solid phases.

It should be noted, however, that under qualitative analysis we will focus only on a few important concepts which include: stoichiometry and non-stoichiometry compounds, ionic equilibria, solution and concentration units, chemical equilibrium, pH, buffer, and buffer capacity, and solutions.

2.2   Stoichiometry

The *law of conservation of mass* can be explained by the fact that atoms are neither created nor destroyed during chemical reactions; rather, atoms are only rearranged during chemical reactions. That is why we are always using atom to write a balanced equation for the chemical reactions. The coefficients that are obtained in writing a balanced overall equation tell us the quantities of the substances involved in the reaction, and therefore tell us the mass relations between all the substances in the reaction. The study of the quantitative nature of chemical formulas and chemical reactions is known as *stoichiometry:* the term is derived from the Greek stoicheion (simplest elements/components) and metric (to measure). The coefficients in a balanced chemical equation are called *stoichiometry coefficients,* and are used to determine relations between the masses of reactants and products of chemical reactions. This is based on the assumption that this is the only reaction occurring and that every atom consumed as a reactant must appear in one of the products. The coefficients of a balanced chemical equation can be interpreted in terms of atoms, molecules or formula units and therefore moles, since there is 6.022×10^{-23} of anything in one mole. The number of moles can easily be converted to grams giving the mass relations for the reactants and products. Let us consider the development of these concepts using the example below:

$$2\,Mg\,(s) \quad + \quad O_2\,(g) \quad \longrightarrow \quad 2\,MgO\,(s) \tag{2.1}$$

2 atoms	1 molecule	$\longrightarrow$	2 formula units
$2\,(6.022 \times 10^{-23})$ atoms	$1\,(6.022 \times 10^{-23})$ molecule	$\longrightarrow$	$2\,(6.022 \times 10^{-23})$ formula units
2 moles	1 mole	$\longrightarrow$	2 moles
48.0 g	64.0 g	$\longrightarrow$	112 g

Obviously, there are several additional points that should be made in connection with this information:

(1). Magnesium and oxygen often react in the same mole proportions to form magnesium oxide, and therefore often react in the same mass proportions, *though always there is the chance of the actual numbers of grams involved varying,* and

(2). There can be a conservation of mass during a chemical reaction without having a conservation of moles.

Suppose you want to know the number of moles and number of grams of MgO that could be prepared from say 6.8 moles of Mg. You could convert moles of Mg to the stoichiometrically equivalent number of moles of MgO and then proceed to convert moles of G_o to grams of G_o. That is, you can follow the logic pattern as follows:

The conversion of moles of Mg to the stoichimetrically equivalent number of moles of G_o requires the use of stoichiometric factor. A *stoichiometric factor* is mole-ratio factor relating moles of one substance appearing in a balanced equation to moles of another substance appearing in the same balanced equation. The stoichiometric factor comes directly from the coefficients in the balanced equation. In the previous case, the desired stoichiometric factor for converting moles of Mg to the stoichiometrically equivalent number of moles of G_o is obtained from the coefficients in the balanced reaction equation. That is why it always important to write correctly balanced equations for chemical reaction (2.1). Therefore, from the latter equation 2 mole Go/2 mol Mg is in a ratio of 1:1. Thus, converting 6.8 mole of Mg to moles of G_o can be accomplished by using equation (2.2):

$$6.8 \text{ mol MgO} = 6.8 \text{ moles } x \frac{2 \text{ mol MgO}}{2 \text{ mol Mg}}$$

(2.2)

and converting 6.8 mol to grams of G_o can be obtained by using expression 2.3:

$$380.8 \text{ } g \text{ } MgO = 6.8 \text{ moles } x \frac{2 \text{ mol MgO}}{2 \text{ mol Mg}} x \frac{56.0 \text{ g MgO}}{1 \text{ mol MgO}}$$

(2.3)

The above conversions are examples of mole-to-mole and mole-to-mass conversions for chemical reaction.

2.2.1   Reactions in Which One Reactant is Present in Limited Supply

In very few cases you will find reactants combined in stoichiometric amounts to conduct chemical reactions. Often, relatively large excess inexpensive reactant is supplied to ensure that a more expensive reactant will be completely converted to products. In such case, the starting amounts of two or more reactants are known, and the challenge is to calculate the maximum amount of product that can be produced. It is important to note that the compound that is completely consumed first is called the *limiting reactant* or *limiting reagent*, because its amount determines or limits the amount of the product to be formed.

The following is the general procedure in determining which reactant is the limiting reactant:

$$aX \quad + \quad bY \longrightarrow \text{products}$$

when mass of both X and Y are given, one should do the following:
(i) calculate the number of moles of X and Y available by converting grams X to moles X, and grams of Y to moles of Y, and
(ii) calculate the value of ratio:

$$\frac{\text{moles of } X \text{ available}}{\text{moles of } Y \text{ available}}$$

by dividing moles of X by moles of Y,
calculate the ratio:

$$\frac{\text{moles of } X \text{ required}}{\text{moles of } Y \text{ required}}$$

by dividing the coefficient **a** to the coefficient **b** of the balanced equation and compare the values of the ratios

$$\frac{\text{moles of } X \text{ available}}{\text{moles of } Y \text{ available}} \text{ to } \frac{\text{moles of } X \text{ required}}{\text{moles of } Y \text{ required}}$$

and determine that X is the limiting reactant when

$$\frac{\text{moles of } X \text{ available}}{\text{moles of } Y \text{ available}} < \frac{\text{moles of } X \text{ required}}{\text{moles of } Y \text{ required}}$$

because there is too much of Y available for X, and that Y is the limiting reactant when

$$\frac{\text{moles of } X \text{ available}}{\text{moles of } Y \text{ available}} > \frac{\text{moles of } X \text{ required}}{\text{moles of } Y \text{ required}}$$

because there is too much X available for Y.

For clarity of stoichiometry reaction equations, usually chemists frequently include information about the physical state of substances in equations: thus, (g), (l), (s), and (aq) refer to gaseous, liquid, solid and aqueous solution states, respectively.
Also expression on mass in chemical units such as:
mole (mol.) i.e.,

$$mole = \frac{g}{fw}$$

Millimole (mmol) i.e.,

$$mmol \ \frac{g}{mfw}$$

Equivalent (eq) i.e.,

$$eq = \frac{g}{eqw}$$

Let us now look at one sample problem, which is a good example of stoichiometry.

Example 2.1

How may grams of Cu_2S can be produced from 9.90 g of CuCl reacting with an excess of H_2S gas?

Solution: Below are the steps to be followed:

Step 1: You must write and balance the equation of the reacting CuCl and H_2S to produce Cu_2S. For this case we usually assume that only one reaction will occur and that all of the reactants are used. In actual fact, a number of reactions may be occurring, the real reaction may not be well known, or all reactants may not necessarily react.

Step 2: Establish the number of moles of the given substance as follows;

$$Mole = \frac{mass \ of \ give \ substance}{Molar \ mass \ of \ given \ substance}$$

$$Mole = \frac{9.9 \ g \ CuCl}{99.0 \ g \ CuCl}$$

Mole = 0.1 mole of CuCl

Step 3: The mole ratio of a given reaction equation of reacting substances should be determined. For convenience, both H_2S and HCl are not considered though are part of the reaction. The amount of Cu_2S produced by 0.1 mole of CuCl can be obtained as follows:

$$= \frac{159 \ x \ 0.1 \ g. \ mole \ Cu_2S}{2 \ mole}$$

$= 7.95$ g of Cu_2S

Step 4: Because we are asked to determine grams of Cu_2S, we should convert the number of moles into grams by using molar mass of Cu_2S.

Thus, 9.90 g CuCl will react with H_2S to produce 7.95 g of Cu_2S.

Practice 2.0
How many grams of silver chloride can be produced from the reaction of 17.0 g of silver nitrate with excess sodium chloride? **Solution:** 14.4 g AgCl

2.2.2   Nonstoichiometric Compounds
There are some inorganic compounds (solids) which have no constant, integral rations of atoms. A good example is ferrous oxide has the composition $FeO_{1.06-1.09}$. The deficiency in iron arises from cation vacancies in the lattice; however, electrical neutrality is maintained by replacement of some Fe^{2+} ions by Fe^{3+}. In some cases some solids may have an excess of cations. Usually this is due to the presence of cations in the interstices between the ions of the normal lattice or from anion vacancies, and also charge balance is attained through substitution of one ion by another of the same sign, but different charge. Nevertheless, nonstoichiometric compounds are occasionally encounted in chemical analysis.

2.3   Ionic Equilibria in Solutions
2.4   Solutions and Concentration Units
Analytical determination often involves working with solutions. Material to be analyzed should be dissolved in appropriate solvent, which may be acid, distilled water or a nonaqueous solvent, for example toluene, tetrahydrofuran (THF), liquid ammonia, hydrogen fluoride, liquid sulfur dioxide, etc. Upon completing dissolution, a variety of operations such as extraction, precipitation, titration or measurement of light absorption is carried out on the resulting solution. It is imperative therefore, for an analytical chemist or anyone concerned with analytical procedures to be aware of some fundamental aspects of solution chemistry.

Let us define the term **solution**. This is defined as a homogeneous mixture of two or more chemical substances. *Homegeneous mixture* implies that the properties of the smallest noticeable portion of the solution are identical to the properties of the whole mixture.

It is apparent that a suspension, such as oil in water, is obviously not homogeneous, and in such a suspension one can easily see particles of oil suspended in water, or *vice versa*. No doubt the properties of oil particles, such as colour, density, and refractive index, are quite different from those of water. At some point a finely divided suspension may look homogeneous to the naked eye, but shows heterogeneity under the microscope. In addition, there are suspensions contain agglomerates consisting of thousands of molecules of matter which are known as colloidal dispersions.

Usually, such colloidal consists of particles of varying sizes from about 1000 Å in a diameter down to 10 Å.

Generally speaking, most of the common compounds we encounter in analytical chemistry are composed of molecules which are less than 10 Å in diameter, though there are numerous naturally occurring compounds with very large molecules (such as proteins), whose molecules may reach into millions. This poses a big challenge in making a distinction between colloidal dispersions and molecular solution based on the size of the particles.

Obviously, a solution in chemistry can be gaseous, liquid or solid. The discussion of solutions in this chapter will only focus on liquid solutions.
In preparing a solution there are two important components, which are *solvent* and *solute*. The *solvent* is the dispersing medium and the *solute* is the dispersing substance. A good example is when we dissolve sodium chloride in water, the salt stands for *solute* and water stands for solvent, irrespective of their relative amounts. For the case where a solution is made up of two or more liquids, the distinction between solute and solvent disappears. Arbitrarily, one of the liquids which is present in large excess is called solvent. It is often desirable to know the amount of solute present per given amount of solvent. The latter ratio is known as *concentration* of a given solution. Concentration values depend on convenient units selected to express the amounts of solute and the solvent. Let us look at some important concentration units as follows:

Percentage by Weight: this is an expression of the ratio of the weight of solute to that of solution. This is often used only for concentrated solutions and when a precise concentration is of little importance. A good example is 100 g of an aqueous 5% calcium hydroxide solution which contains 5 g of $Ca(OH)_2$ and 95 g of water.

Percentage by Volume: this is the ratio of the volume of solute to that of the solution in another concentration unit that is used. It is commonly used in liquid-liquid solutions. It should be clear that this unit is somehow ambiguous because, (a) the volume may change with temperature, and (b) the volumes of the mixed liquids are not always additive: there is a chance of expansion or shrinking of the solution up on mixing. A 10% solution by volume of alcohol in water implies that the solution has been made up by the addition of 10 volumes of alcohol to 90 volumes of water at given temperature.

Molarity (M): this is the number of moles of solute in one litre of solution. As we have seen, the volume of solution depends on the room temperature, and for correct analytical work, a solution with a certain molarity at one temperature should be used at this temperature only. On the other hand, fluctuation of 1-2 ºC does not appreciably affect the molarity in most analytical work. For example, if an aqueous solution is 1.0000 M at 25 ºC, it would be 0.9995 at 27 ºC. The difference of 0.5 parts per 1000 is relatively small to significantly affect most experimental results.

There are some texts where the term *formal* is used instead of molar, just to indicate the number of formula weights of a given substance per litre of solution. Nevertheless, the term *'molar'* is commonly used with little need for change.

Molality (m) is the number of moles of solute per 1000 g of solvent. Obviously, the concentration unit is independent of temperature fluctuations and is commonly used in physical chemistry. The molality of an aqueous solution is approximately equal to the molarity of the solution when the molarity is less than 0.01 M, because, under these conditions, the volume of solution plus volume of one kilogram of water is essentially equal to the volume of one litre of solution.

Mole-fraction is the ratio of the number of moles of the solute to the total number of moles of all of the substance present in solution. Suppose you have a mixture of two substances, C and D, the mole fraction is expressed as follows:

$$X_C = \frac{n_C}{n_C + n_D} \tag{2.4}$$

where X_C is the mole fraction of C in a solution containing n_C moles of C and n_D moles of D. Since the sum of the mole fractions is necessarily equal to one for any mixture,

$$X_C + X_D = 1.00 \tag{2.5}$$

The *weight percent* of component C is equal to the mass of C divided by the total mass of solution times 100.

$$wt.\% \; C = \frac{gram\ of\ C}{grams\ of\ C + grams\ of\ D} \; x\ 100 \tag{2.6}$$

Think of a solution composing one solute C and water, expression 2.6 becomes

$$wt.\% \; C = \frac{gram\ of\ C}{grams\ of\ C + grams\ of\ water} \; x\ 100 \tag{2.7}$$

The *parts per million* of component C, ppm C, is defined as the milligrams of component per litre of solution for aqueous solutions.

$$ppm\ C = \frac{milligrams\ of\ C}{Litre\ of\ solution}$$

$$(2.8)$$

It is assumed that student is well acquainted with concentration units and can convert them readily.

Example 2.2
An aqueous solution contains 10% by weight of sodium chloride. Calculate (a) its molarity, and (b) the mole fraction of the solute present.

Solution
(a) The solution contains 10 g of NaCl per 90 g of water; therefore, in 1000 g of water there are 111.1 g of NaCl or 1.90 moles NaCl. The solution is 1.90 *molar* (M).

$$\frac{10\ x1000}{90} = 111.1\ g \quad \text{or}$$

$$1.90\ moles = \frac{111.1}{58.5}$$

(b) In 1000 g of water there is

$$55.50\ mole = \frac{1000}{18.02}$$ of water. This means we have a mixture of 55.50 moles of water and 1.90 moles of NaCl.

$$X_{NaCl} = \frac{1.90}{1.90 + 55.50} = \frac{1.90}{57.4} = 0.033$$

Practice 2.1
What is the molarity of a solution prepared by dissolving 17.5 g of $Na_2CO_3.10\ H_2O$ in enough water at at room temperature to make a 1 litre of solution? **Solution** is 0.0612 *molar* (M).

Practice 2.2
Concentrated phosphoric acid containing 85 per cent H_3PO_4 has a density of 1.69 gm/mL. Calculate the number of grams and moles of H_3PO_4 in 1 litre of acid. What is the molarity of the acid? **Solution** is 1437 g; 14.7 moles.

2.5   Chemical Equilibrium
The study of ionic equilibria in analytical chemistry cannot be overemphasized. Because whenever we apply chemical reaction for analytical determination, we prefer most the reaction which is *fast* and *complete.* In precipitation, for example, we would like the precipitate to be insoluble, so that all constituents would be found in the precipitate. Also in

filtration it is desirable that each equivalent of the reagent reacts with titrated substance to completion. In many cases we do not approach such ideal conditions. In short, every chemical reaction in a solution reaches an equilibrium state short of completion. That is why the study of equilibrium conditions is of paramount importance. This section assumes that a student at that level has covered the factors affecting chemical equilibrium and we will discuss them briefly in this chapter in section 2.5.

$$(2.9)$$

As the reaction proceeds, A and B will be used up and compounds C and D will be formed. This depends on the chemical reaction involved and factors such as temperature, pressure, catalyst (which vary from a minute fraction of a second to many centuries), the system when it reaches the equilibrium (the mixture remains the same). Because reactions do not get to completion that is why you cannot find only substance C and D present, and there will be always some unreacted A and B remaining in solution. The cause of incompleteness of reactions in solutions is due to the fact that they are reversible. It is important to remember that if one of the reaction products is removed continuously from the reaction medium, the reaction ceases to be *reversible* and will go to completion.

For example,

$$(2.10)$$

If the above reaction is carried out in an open beaker, obviously H_2 (g) gas will escape into the atmosphere. Under such condition the reaction will go to completion. The rate of a chemical reaction, in general, depends on many factors. Some important ones are: temperature, pressure, nature of solvent (if any), catalyst, concentrations and chemical affinity of the reactants. In addition, in a given chemical reaction, at constant temperature, pressure, etc., the rate of reaction always depends on the amount of the concentration of the reactants (species). Such dependency was established in 1864 by Guldberg and Waage. The formulation since then is known as the *Law of mass action* and is stated as, "The rate of a reaction is proportional to the active masses of the reacting substances, each raised to the power equal to the number of molecule appearing in the balanced equation."

For the equilibrium reaction (2.9), the rate of the forward reaction at any specific time is proportional to the product of the concentrations of A and B existing at that time. Therefore,

$V_f \sim$ (A) (B) or $V_f = K_f$ (A) (B)

K_f is proportionality constant; V_f is rate of forward reaction

For the reverse reaction

$V_r = K_r (C) (D)$

K_r is the proportionality constant; V_r is rate of reverse reaction

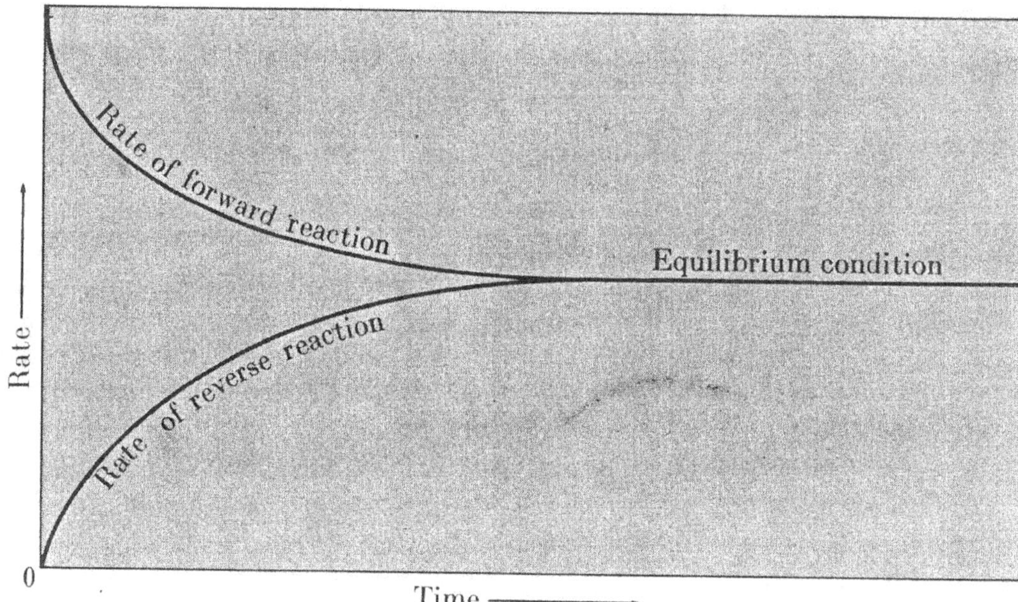

Figure 2.1: Equilibrium Condition

At equilibrium condition in Figure 2.1, the rates approach each other and finally the forward and reverse reactions become equal and we have: K_f (A) (B) = K_r (C) (D)

$$\frac{K_f}{K_r} = \frac{(C)(D)}{(A)(B)} = K$$

K is equilibrium constant for the reaction. Since the reaction is reversible, we can express it as follows:

The equilibrium constant then becomes

$$\frac{(A)(B)}{(C)(D)} = K'$$

with minimum confusion it can be easily seen as follows:

$$\frac{1}{K} = K'$$

Generally speaking, all chemical reactions are *reversible*. Due to that fact, the state of dynamic equilibrium is reached in which the forward and reverse reactions continue to occur, but they take place at equal rate (Figure 2.1). This fact reveals that the nature of the equilibrium state for a specific chemical reaction is the same at any given temperature, no matter what the direction of the approach. Indeed, this means we can start with either reactants or products and be sure to find the same distribution of reactants and products at equilibrium, as long as we are using the same amount of material.

An equilibrium constant expression 2.12 relates the concentrations of reactants and products at equilibrium at a given temperatures to a numerical constant. For the reaction 2.11 the equilibrium concentrations of reactants and products are related by the equilibrium constant expression 2.12. Product concentrations are always placed in the *numerator* and reactant concentrations in the *denominator*. Furthermore, the concentration of each substance is raised to the power of its stoichiometric coefficient in the balanced equation. The value of K_c often depends on the particular reaction and on the temperature. The K value usually tells us all about the extent of reaction that has occurred when equilibrium has been reached.

$$(2.11)$$

$$K_c = \frac{(c)^2 (D)^2}{(A)^2 (B)}$$

$$(2.12)$$

and for

$$(2.13)$$

$$K = \frac{(B)(C)}{(A)^2}$$

Example 2.3
Why does the concentration of the reactants and the products appear in the equilibrium constant expression?

Solution
The reason is because the rate of reaction is proportional to the total number of collisions per unit time. In order for a reaction to occur between two molecules, they obviously have to collide with each other, lthough not every collision results in a reaction (only high energy collisions favour a reaction). For the purpose of determination of K, the reaction concentration of reactants and product at equilibrium in the reaction mixture must be known.

Example 2.4

Suppose we have the following reaction with 1 molar of both A and B. At equilibrium it was found that the concentration of C is 0.25 molar. We know that for every mole of C produced in the reaction, an equivalent amount of D is also produced. If C is 0.25 molar, concentration of D will likewise be 0.25 molar. Therefore, the concentration of A and B at equilibrium is 1.0 - 0.25 = 0.75 molar.

Solution

Thus, K can be calculated as shown below

$$K = \frac{(C)(D)}{(A)(B)} = \frac{0.25 \; x \; 0.25}{0.75 \; x \; 0.75}$$

$K = 0.111$

Example 2.5

Consider the following dissociation reaction

$$K = \frac{(C)(B)}{(A)} \tag{2.14}$$

Solution

We are given an initial concentration of A as 1.0 molar, and α is degree of dissociation, which is fraction of A that underwent dissociation under equilibrium conditions. The concentration of both B and C will be α molar concentration, respectively.

Substituting the values into expression 2.13 we get the following:

$$K = \frac{(\alpha)(\alpha)}{(1-\alpha)} = \frac{\alpha^2}{(1-\alpha)}$$

If $\alpha = 0.1$, that is, if we have dissociation of 10%, the equation becomes:

$$K = \frac{(0.1)^2}{(1 - 0.1)} = \frac{0.01}{0.9}$$

$K = 0.011$

In general, if the original concentration of substance A is designated as C, then at equilibrium A will be $(1- \alpha)$C and the equilibrium concentration of B and C will both be equal to αC. Substituting this in expression 2.14 results in the following:

$$K = \frac{(\alpha C)(\alpha C)}{(1-\alpha)C} = \frac{\alpha^2 C^2}{(1-\alpha)C} \quad \text{or}$$

$$K = \frac{\alpha^2 C}{(1-\alpha)}$$

The above equation is known as the **Ostwald Dilution Law.**

From the above expression α must lie between 0 and 1. If α is 0, $K = 0$ and there is zero dissociation, also when α is 1, $K = \infty$ and the dissociation is complete.

Example 2.6

A 0.100 M solution of a weak acid HA is 2.5% dissociated. Calculate (a) the dissociation constant, and (b) the percent dissociation of HA in a 0.001 M.
Solution
(a) Given that $\alpha = 0.025$, we can calculate the dissociation constant using the **Ostwald Dilution Law.**

$$K = \frac{0.10 \; x \; (0.025)^2}{(1 - 0.025)} = 6.41 \; x 10^{-5}$$

(b) Upon reversing the above procedure, and given that $C = 0.0010$, one can calculate the value of α from the same equation.

$$\frac{0.0010 \; x \; (\alpha)^2}{(1 - \alpha)} = 6.41 \; x 10^{-5}$$

or

$$\frac{0.0010 \; x \; (\alpha)^2}{(1 - \alpha)} = 6.41 \; x 10^{-5}$$

$$\alpha^2 + 6.41 \; x \; 10^{-2} \; \alpha - 6.41 \; x \; 10^{-2} = 0$$

A student can solve the quadratic equation, and get $\alpha = 0.223$
Therefore, in a 0.0010 M solution the acid is dissociated to the extent of 22.3%. The complete solution of the quadratic equation will also yield a negative root of α, but, of course, it does not have any physical significance and it is not often considered.

Writing equilibrium concentration expression for reaction involving solids and water: this attempt is always easier because the molar concentration of a solid is fixed by its density and molar mass expression 2.15 and

$$\frac{g}{mL} \cdot \frac{10^3 \ mL}{1 \ L} \cdot \frac{1 \ mole}{M \ g} = \frac{moles}{L}$$

$$(2.15)$$

cannot be affected by the reaction of certain amount of solid or addition of some solid. Indeed, the equilibrium concentrations of other reactants and products are also not changed by the amount of solid present in a reaction, as long as some solids are present at equilibrium. That is why chemists do not include the concentrations of any solids in equilibrium constant expression 2.18.

Consider the concentrations of solids as already being included in the value of K for the reaction. For example

$$K' = \frac{[SO_2 \ (g)]}{[S \ (s)]^{1/8} \ [O_2 \ (g)]}$$

$$(2.17)$$

so,

$$K' \cdot [S \ (s)]^{\frac{1}{8}} = K = \frac{[SO_2 \ (g)]}{[O_2 \ (g)]}$$

$$(2.18)$$

The molar concentration is also fixed by its density and molar mass, and has the value of 55.6 M at room temperature: see expression 2.19.

$$\frac{g}{mL} \cdot \frac{10^3 \ mL}{1 \ L} \cdot \frac{1 \ mole}{18.0 \ g} = 55.6 \ \frac{mol}{L}$$

$$(2.19)$$

Due to this reason, the concentration of water is always not changed by chemical reactions occurring in aqueous solutions and it is therefore not included in equilibrium constant expressions for reactions occurring in aqueous solutions.

2.5.1 The Meaning of Equilibrium Constant

The value of equilibrium constant implies whether a reaction is reactant or product favoured. A large value of K >> 1 means the ratio of the product concentrations over reactant concentrations, each raised to the power of their stoichiometric coefficient, is large, and the reaction is referred to as *product* favoured. Conversely, a small value of K << 1 means the concentrations of the reactants are large compared to those of the products and the reaction is *reactant* favoured. Once K is known it can be used with initial concentrations and the stoichiometry of the reaction to calculate the concentrations of reactants and product at equilibrium.

2.6　The Reaction Quotient, Q

The reaction quotient expresses the concentrations of reactants and products at any point in a reaction at a given temperature. From the general reaction

The reaction quotient expression is:

$$Reaction\ quotient = Q_c = K = \frac{[C]^c[D]^d}{[A]^a[B]^b}$$

(2.21)

It apparent that the form of a reaction quotient expression 2.21 is identical to the corresponding equilibrium constant expression 2.12, but Q_c differs from K_c in that concentrations used in the reaction quotient expression are not *necessarily equilibrium concentrations*. It should be clearly noted that K_c is a special value of Q_c pertaining to equilibrium conditions.

The following is the existing relationship between the reaction quotient Q_c, and the equilibrium constant K_c.

When

- $Q < K$, the ratio of product concentrations over the reactant concentrations is too small, and the reactants must be converted to products until $Q = K$.

- $Q > K$, the ratio of product concentrations over reactant concentrations is too large, and products converted to reactants until $Q = K$.

2.7　Disturbing a Chemical Equilibrium

Le Chatelier's Principle

There are three important known ways in which a chemical reaction system at equilibrium can be disturbed: (i) by a change in concentration of reactant or product, (ii) by change in temperature or (iii) by change in volume of the container. The outcomes of such changes are well predicted by the Le Chaterlier's principle, which state: A *change* in any of the factors that determine the equilibrium conditions of a system causes the system to change in a manner that will counteract the effect of the change.

2.7.1　Effect of Concentration Changes on Equilibrium

Upon changing the concentration of a reactant or product from its equilibrium value at a given temperature, the reaction will shift to a new equilibrium position for which the reaction quotient is still equal to K. This can be summarized as follows:

- Addition of a reactant or removal of product causes Q to temporarily be less than K and the reaction to proceed in the reactant side to re-establish equilibrium, and

- Removal of a reactant or addition of product causes Q to temporarily be greater than K and the reaction to proceed in the reactant side to re-establish equilibrium.

2.7.2   Effect of Temperature Changes on Equilibrium

Upon changing (by rising) the temperature of a system at equilibrium by adding heat energy, the system reacts in the direction that absorbs heat to restore a state of equilibrium. This tells us that an increase in temperature always causes the system to shift in the endothermic direction to achieve the new state of equilibrium, because heat can be thought of as a reactant in an endothermic change. Conversely, lowering the temperature of a system at equilibrium causes a shift in the heat releasing or exothermic direction. It should be clearly noted that *only change in temperature* causes a change in K. If the reaction is endothermic in a forward direction, that direction is enhanced by an increase in temperature. This means the concentrations of the products of an endothermic reaction increase and those of the reactants decrease with an increase in temperature, and the value of K, which is a reflection of the ratio of product concentrations over reactant concentrations, *increases with increasing temperature for endothermic reaction. Conversely, the value of K for an exothermic reaction decreases with increasing temperature,* because the rise in temperature opposes the tendency of exothermic reaction to give off energy as heat.

2.7.3   The Effect of Volume Change on Gas Phase Equilibrium

When the volume of the container containing a gaseous system at equilibrium is decreased, there is always a pressure increase that is counteracted by a shift to the reaction side that contains the *fewer number of moles of molecules* of gases. On the other hand, an increase in the volume of the container causes a pressure drop that is counteracted by a shift to the reaction direction that contains the greater number of moles or molecules of gases.

2.8   The pH Scale

Some time in the past there was a need to know a more convenient way to describe hydronium, H_3O^+, and hydroxide, OH^- ion concentrations. In 1909, Sørensen proposed the term pH to refer to the potential of hydrogen ion concentrations. He defined it as the negative of the logarithm of $[H_3O^+]$. It is important to note that the concentration of hydronium or hydroxide ions in aqueous solution is always quite small. Although scientific notation has helped in expressing these very small numbers, nevertheless there are still some challenges.

Furthermore, Sørensen defined pH as negative of logarithm of $[H^+]$ and can be restated in terms of $[H_3O^+]$. Indeed, it should be clear that pH is an important and convenient way for expressing $[H_3O^+]$ in aqueous solution. Think of an acidic solution $[H_3O^+]$ to be equal to 10^{-7} M. A simplified scientific notion obviously looks better than using a string of non-significant zeros (i.e., 0.0000001 M), but when expressed as pH, the number is simply 7.0. The pH scale represents the negative exponent of 10 as a positive number. The exponent of 10 in a number is the numbers common logarithm or simply, log.

$pH = -\log [H_3O^+]$

When expressing pH, usually the number in the right of the decimal place must have the same number of significant figures as the original number,

$$H_3O^+ = 1.00 \; x \; 10 \; x \; 10^{-5} \, M \qquad pH = 4.000$$

3 significant figures   3 places to the right of the decimal

that is, if

$pOH = -\log [OH^-]$ this expression is valid, though is not commonly used. The relationship between pH and pOH is usually derived from the ionic product of water.

$[H_3O^+][OH^-] = 1.0 \; x \; 10^{-14}$ we will see it later.

Example 2.7

Autoprotolysis of water: H_2O

$$(2.22)$$

$$Kw = [H^+][OH^-] \qquad\qquad (2.23)$$

whereas; w - stands for water

$[H_3O^+][OH^-] = 1.01 \; x \; 10^{-14}$

The value of Kw at room temperature 25 °C is $1.01 \; x \; 10^{-14}$. Calculate the concentration of H^+ and OH^- in pure water at 25 °C.

Solution

Stoichiometry of reaction in expression 2.22 tells us that H^+ and OH^- are produced in 1:1 molar ratio. Their concentrations are assumed to be equal. Let each concentration be x, and we can write.

$Kw = 1.0 \; x \; 10^{-14} = [H^+][OH^-] = [x][x] = Kw = x^2 = 1.0 \; x \; 10^{-14} \rightarrow x = 10^{-7} \, M$

The concentration of H^+ and OH^- are $1.0 \; x \; 10^{-7} \, M$ in pure water.

Example 2.8

What is the concentration of OH^- if $[H^+] = 10^{-3} \, M$? Assume the temperature is 25 °C.

Solution: Putting $[H^+] = 1.0 \; x \; 10^{-3} \, M$ into Kw expression 2.33.

(iii) $Kw = 1.0 \; x \; 10^{-14} = (1.0 \; x \; 10^{-3})(OH^-) \rightarrow [OH^-] = 10^{-11} M$. As the concentration of H^+ increases, the concentration of OH^- necessarily decreases, and vice versa. The pH of pure water at 25 °C is

$pH = -\log [H^+]$

$pH = -\log (1.0 \; x \; 10^{-7})$

$pH = 7.00$

The useful relationship between concentrations of H^+ and OH^- is

pH + pOH = -log Kw = 14.00 at 25 $^\circ$C

Kw = [H^+] [OH^-]

log Kw = log [H^+] + log [OH^-]

log Kw = log [H^+] + log [OH^-]

By introducing negative (-) to the above expression the following is obtained

-log Kw = -log[H^+] – log [OH^-]

$$pKw = pH + pOH \qquad\qquad (2.24)$$

14 = pH + pOH at 25 $^\circ$C

If pH = 3.58; therefore   pOH = 14 – 3.58 = 10.42

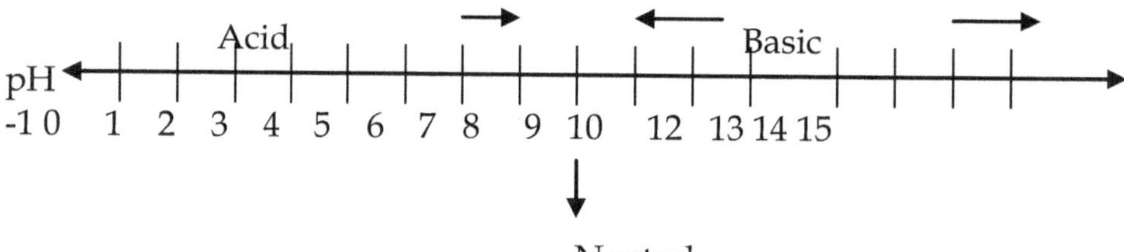

Although pH generally falls in the range of 0 to 14, these are not the limits of pH.

Example 2.9

Determine (a) [H_3O^+] in the rainwater if its pH measured 4.35 (b) [OH^-] in the ammonia if its pH measured 11.28

Solution

(a) pH = - log [H_3O^+]

   log [H_3O^+] = - pH = 14.5 x 10^{-5} M

Establish pOH from expression 2.24

pOH = 14 – pH = 14.00 – 11.28 = 2.72

Now use the definition of pOH = - log [OH^-]

log [OH^-] = - pOH = 2.72

[OH^-] = $10^{-2.72}$ = 1.9 x 10^{-3} M

Example 2.10

Calculate the concentration in aqueous solution of a strong acid. Calculate [H_3O^+], [Cl^-], and [OH^-] in 0.015 M HCl (aq). Assume that HCl is completely ionized and is the only source of [H_3O^+] = 0.015 M. Also one Cl^- ion is produced for every H_3O^+ ion, and

[Cl^-] = [H_3O^+] = 0.0015 M

Solution
To calculate OH- we must use the following facts
1. All the OH- is derived from the self-ionization of water, i.e.,

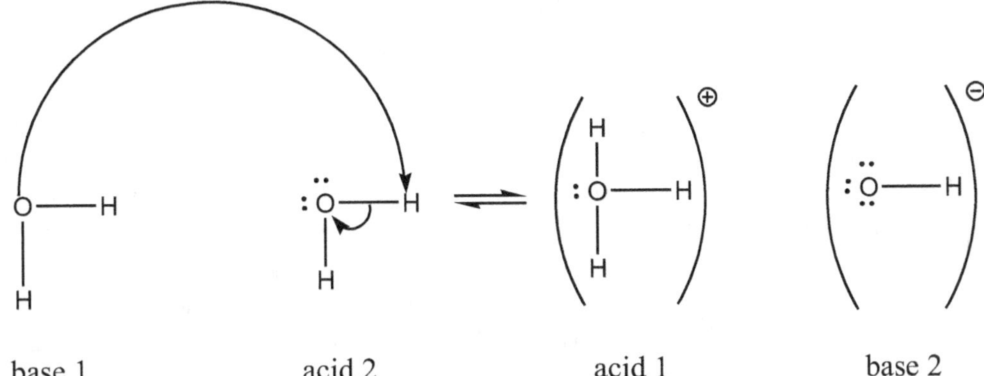

base 1   acid 2   acid 1   base 2

2. [OH-] and [H$_3$O$^+$] must have values consistent with Kw for water.

Kw = [H$_3$O$^+$] [OH-] = 10^{-14}

Kw = (0.015) (OH-) = 1.0 x 10^{-14}

$$[OH^-] = \frac{1.0\ (10^{-14}}{0.015} = 0.67\ x10^{-12}$$

[OH-] = 6.7 x 10^{-13} M

Practice 2.3
A 0.100 M solution of hydrozoic acid (N$_3$H) has pH = 2.83. Find pKa for this acid
Solution pKa = 4.65

2.9   Definition of Buffer Solution

The word *'buffer'* conceptually is used to mean "absorbing shock"; for example, a car bumper works as buffer for the passengers by absorbing the energy in case of impact. A buffer solution has a similar effect on the pH of a solution. Buffer is a mixture of a weak acid and its conjugated or weak base and its conjugate acid. A buffer functions in the presence of substance in solution, which is available to react with weak acids and weak bases, respectively. Think of hypothetical reaction between weak acid HA and OH: two possible reactions are obtained - see expressions 2.25 and 2.26, respectively.

(2.25)

(2.26)

The two reactions above are basically complete. This tells us that with a solution with significant concentration of both weak acid and its conjugate weak base, we have a solution that reacts with either OH$^-$ or H$_3$O$^+$. It is absolutely wrong to assume that just the weak base alone would function as buffer because the ionization produces some of its ions. Nevertheless, the amounts of ionized ions are significantly small so we still need additional anion concentration from another source such as a salt.

A **buffered solution** *resists changes in pH caused by the addition of limited amounts of a strong acid or a strong base.* Some commercial products indicate the importance of buffered solutions in items like hair shampoo and aspirin. For human beings such a phenomenon is very important because many reactions in our body (including those that give us life energy), must take place in a controlled pH environment. A good example showing buffer function is in our body cells. This is H$_2$PO$_4^-$ — HPO$_4^{2-}$ buffer system. To this system the dihydrogen phosphate ion (H$_2$PO$_4^-$) acts as acid specie; that is why it is reacted with the base in expression 2.27, and the hydrogen phosphate ion (HPO$_4^{2-}$) ion acts as a base and is reacted with acid in expression 2.28. The two ions are removing OH$^-$ and H$_3$O$^+$ ions once they are produced in our body cells as is shown in expressions 2.27 and 2.28.

$$(2.27)$$

$$(2.28)$$

Another example where buffer solution shows its important function is when it maintains our pH of blood at 7.40, just barely alkaline. It is amazing to note that enzymes run our body in a very narrow pH range. A small deviation of 0.20 pH units either way can cause *coma* or even become *fatal*. The drop in pH below 7.20 results in a condition called *acidosis*, and a rise in the pH to above 7.60 usually the opposite condition called *alkalosis* emerges. No doubt either situation will upset our blood chemistry seriously. On the other hand, an amazing thing is how our blood maintains almost constant pH despite the fact that both bases and acids find their way into the blood stream. Such a constant in blood pH is a result of existing buffering system.

The more important system maintaining blood pH is carbonic acid-bicarbonate buffer which is represented by the expression 2.29. At any moment carbonic acid (H$_2$CO$_3$) part of the buffer will react with OH$^-$ that enters the blood stream in expression 2.30, and more bicarbonate will be formed, which will be excreted by the kidneys. *Hyperventilation*, the process of breathing too fast, usually causes too much CO$_2$ to be exhaled and results in the blood pH to rise above 7.40. Most of the time physical exercise produces extra H$_3$O$^+$ from lactic acid formed in the muscles. The produced H$_3$O$^+$ will react with bicarbonate as you can see expression 2.31 to produce carbonic acid.

The latter by-product is transported to the lungs, where it is decomposed to carbon dioxide gas and water, which is afterwards exhaled. *The take home message is that the concentration of bicarbonate is almost 10 times the concentration of carbonic acid; this implies that our body is more protected against extra acids than extra bases.*

$$(2.29)$$

$$(2.30)$$

$$(2.31)$$

Generally speaking, a buffer solution requires the following: (1) it should be able to neutralize acids, which means it should have OH⁻ ions, and (2) it should be able to neutralize bases, which means it should have H_3O^+ ions. Nevertheless, the two should not neutralize each other. Therefore, a buffer is usually composed of approximately equal quantities of the members of a conjugate pair: a weak acid and its weak conjugate base or a weak base and its weak conjugate acid.

2.9.1 What You Mix is What You Get

If you mix A moles of a weak acid and B moles of its conjugate base, the moles of acid remain close to A and the moles of base remain close to B. Often, very little reaction occurs to change either concentration. This can be easily understood by looking at Ka and Kb values in terms of Le Chatelier's principle: see section 2.6.

Example 2.12

Think of an acid with pKa = 4.00 and its conjugate base with pKb = 10.00. Calculate the fraction of acid that dissociate in a 0.100 M solution of HA.

Solution

$$Ka = \frac{\alpha^2}{F-\alpha} = 1.0 \times 10^{-4}$$

$\alpha = 3.1 \times 10^{-3}$ M

$$Fraction\ of\ Dissociation = \frac{[HA]}{[A] + [HA]} = \frac{\alpha}{F} = 0.031$$

Only 3.1% of acid has dissociated under these conditions.

In a solution containing 0.100 mol of A⁻ dissolved in 1.00 litre of water, the extent of reaction of A⁻ with water is even smaller. It should always be remembered that F stands for formal concentration of HA.

$$A^- \ + \ H_2O \ \underset{}{\overset{K_b}{\rightleftharpoons}} \ HA \ + \ OH^- \qquad pK_b = 10.00$$

$$0.100 - \alpha \qquad\qquad \alpha \qquad\qquad \alpha$$

$$Kb = \frac{\alpha^2}{F-\alpha} = 1.0 \ x \ 10^{-10}$$

$\alpha = 3.2 \ x \ 10^{-6}$ M

$$Fraction \ of \ Association = \frac{[HA]}{[A]+[HA]} = \frac{\alpha}{F} = \frac{3.2 \ x \ 10^{-6} M}{0.100 \ M} = 3.2 \ x \ 10^{-5}$$

HA dissociates very little, and the Le Chatelier's principle remands us that adding extra A- to the solution will make the HA dissociate even less. Similarly, A- does not react very much with water, and adding extra HA makes A- react even less. If 0.050 moles of A- plus 0.036 mol of HA are added to water, there will be close to 0.050 mole of A- and close to 0.036 mol of HA in the solution at equilibrium.

The *Henderson –Hasselbalch Equation*
This is the central equation for buffers and is derived from the rearrangement of the Ka equilibrium expression.

$$Ka = \frac{[H^+][A^-]}{[HA]}$$

$$Ka = log\left(\frac{[H^+][A^-]}{[HA]}\right)$$

$$logKa = log \ [H^+] + \left(\frac{[A^-]}{[HA]}\right) \tag{2.32}$$

Multiply expression 2.32 by -1 and rearrange to isolate $- log \ [H^+]$

$$-logKa = -log \ [H^+] - log\left(\frac{[A^-]}{[HA]}\right) \tag{2.33}$$

Substitute expressions 2.34 into equation 2.32 and expression 2.35 is obtained.

$$\tag{2.34}$$

$$pH = pKa + log\left(\frac{[A^-]}{[HA]}\right)$$

$$\tag{2.35}$$

This is *Henderson–Hasselbalch Equation* for an acid, and is valid when the ratio of [conjugate base]/[acid] is no larger than ≈ 10/1 and no smaller than

$\approx 1/10$ because these are the limits in which there can be buffering action. Indeed, the best buffering action is obtained when the ratio approximately equals to 1, and the pH is, therefore, approximately equal to the pKa of the acid.

The expression 2.35 tells us that the pH of a solution of a weak acid and its conjugate base is controlled primarily by the strength of the acid, as expressed by its pKa. The *fine control* of the pH is then given by the relative amounts of acid and conjugate base.

If

- The amount of acid is less than the amount of conjugate base, the pH is less than pKa, because log ([conjugate base]/[acid]) is negative.
- The amount of acid is equal to the amount of conjugate base, the pH is equal to pKa, because the logarithim of 1 (log 1) equal to zero (0).
- The amount of acid is greater than the amount of conjugate base, the pH is greater than the pKa, because log ([conjugate base]/[acid]) is positive.

If a solution is prepared from the weak base, B, and its conjugate acid, the analogous expression *Henderson–Hasselbalch Equation* for base is obtained as in expression 2.36 below.

$$BH^+ \xrightleftharpoons{\quad K_a = K_w/K_a \quad} B \quad + \quad H^+$$

$$pH = pKa + \left(\frac{[B]}{[BH^+]} \right) \tag{2.36}$$

where pKa stands for the acid dissociation constant of the weak acid BH^+. The important features of expressions 2.32 and 2.33 are that: (1) the base (A⁻ or B⁻) appears in the numerator and (2) pKa applies to the acid in the denominator.

When $[A^-] = [HA]$; $pH = pK_a$
In expression 2.34 the logarithm term is zero, (0) because log 1 = 0 therefore, when $[A^-] = [HA]$, pH = pKa.

Example 2.12
Sodium hypochlorate (NaOCl, the active ingredient of bleach) was dissolved in solution buffered to pH = 6.20. Find the ratio of
$$\frac{[OCl^-]}{[HOCl]}$$
in this solution. Given that pKa for, HOCl = 7.53

Solution

$$pH = pKa + log\left(\frac{[OCl^-]}{[HOCl]}\right)$$

$$pH - pKa = log\left(\frac{[OCl^-]}{[HOCl]}\right)$$

$$-1.33 = log\left(\frac{[OCl^-]}{[HOCl]}\right)$$

$$10^{-1.33} = 10^{log\left(\frac{[OCl^-]}{[HOCl]}\right)}$$

$$\frac{[OCl^-]}{[HOCl]} = 0.047$$

$$6.20 = 7.53 + \left(\frac{[OCl^-]}{[HOCl]}\right)$$

At this point a student should be able to use a calculator to solve the above logarithm problem as we have seen in *Chapter 1*. It is important to bear in mind that calculating the ratio of [OCl]/ HOCl] requires only pH and pKa and it does not matter what else is in [HOCl] or how much NaOCl was added, or what the volume of the solution is.

2.9.2 Buffer Preparations

There are two basic requirements for a buffer solution:
- it must control the pH at the desired value, so expression 2.37 tells us the weak acid selected must have a pKa that is within one unit of the desired pH, and
- it must have sufficient acid and base forms present to react with added strong acid or base and resist change in pH, so it is important to note that buffer solutions often have 0.1 to 1.0 M concentrations of weak acid or conjugate base or weak base and conjugate acid.

$$pH\ range\ for\ buffering = pKa \pm 1 \tag{2.37}$$

Nevertheless, it should be remembered that any buffer loses its buffering capacity if too much of strong base or strong acid is added to it. The preparation starts with measuring the amount of either a weak acid (HA) or a weak base (B). Then OH- is added to HA to make a mixture of HA and A- (a buffer) or H+ is added to B to make a mixture of B and BH+ (a buffer).

Example 2.13

How many millilitrers of 0.500 M NaOH should be added to 10.0 g of this hydrochroride (BH+) to give a pH of 7.60 in a final volume of 250 mL? Given pKa for BH+, the conjugate acid of tris is 8.07

The number of moles of this hydrochloride in 10.0 g is

$$0.0635 = \frac{10.0\ g}{157.60\ g.mol^{-1}}$$

In solving this problem let us use the table kind of as follows:

From Henderson-Hesselbalch equation 2.36 we can find α, because we know both pH and pKa.

$$pH = pKa + \left(\frac{[mol\ B]}{[mol\ BH^+]}\right)$$

$$7.60 = 8.07 + \left(\frac{\alpha}{0.0635 - \alpha}\right)$$

$$-0.47 = \left(\frac{\alpha}{0.0635 - \alpha}\right)$$

$$10^{-0.47} = 10^{log\left(\frac{\alpha}{0.0635 - \alpha}\right)}$$

$$0.339 = \frac{\alpha}{0.635 - \alpha}$$

$$\alpha = 0.0161$$

This many moles of NaOH is contained in

$$0.0322\ Litre = \frac{0.0161}{0.50\ mol.l^{-1}}$$

The above calculation directs us to mix 32.2 mL of 0.500 M NaOH with 10.0 g of trishydrochloride to give a pH = 7.60

2.9.3 Buffer Capacity

Buffer capacity is a measurement which reveals how well a solution can resist pH changes when a small amount of acid or base is added. This means the greater the buffer capacity, the less pH changes when small amounts of H^+ or OH^- are added. Indeed, buffer capacity is considered maximum when pH = pK_a. One has to be careful in choosing a buffer for an experiment by choosing that with pK_a values close to the desired pH. Usually, the useful pH range of buffer is considered to be $pK_a \pm 1$ pH unit. This implies any value slightly outside that range has less weak acid or weak base characteristics to react with added acid or base. Buffer capacity often increases with increasing concentration of the buffer. For good pH maintenance, it is advisable therefore to use enough buffer solution to react with the quantity of acid or base expected.

2.10 Problems

2.10.1. A solution of 100 mL volume contains 0.2083 g of $BaCl_2$.
    (a) How many moles of $BaCl_2$, Ba^{2+}, and Cl^- are present?
    (b) What volume of 0.100 M silver nitrate solution is required for the precipitation of the chloride?

2.10.2. At 25 °C an aqueous solution containing 219.6 g of HCl per litre has a specific gravity of 1.0980. Calculate (a) the molarity, (c) the molality and the percentage by weight of HCl in the solution.

2.10.3. What volume of concentrated hydrochloric acid having a density of 1.200 and containing 39.11 per cent HCl is required to prepare (a) 1 litre of acid containing 20.01 per cent HCl (density 1.100) (b) 1 litre of 6.00 M acid?

2.10.4. (a) What is pH and the fraction of dissociation of a 0.100 M solution of weak acid HA with $Ka = 1.00 \times 10^{-5}$? (b) 0.045 M solution of HA has a pH 0.278. Find pKa for HA (c) If A 0.045 M solution of HA is 0.60% dissociated. Find pKa for HA.

2.10.5. The concentration of human being blood is 3.5×10^{-8} M. (a) What is the pH of the blood? (b) Find the concentration in the blood.

2.10.6. (a) What is the pH of a solution prepared by dissolving 10.0 g of tris plus 10.0 tris hydrochloride in 0.250 litre of water? (b) What will the pH be if 10.5 mL of 0.500 M $HClO_4$ are added to (a)? (c) What will the pH be if 10.5 mL of 0.500 M NaOH are added to (a)?

2.10.7. Calculate the pH of each of the following solutions:
    (a) A 0.10 M solution of NaOH
    (b) A 0.0050 M solution of slaked lime $[Ca(OH)_2]$
    (c) A solution prepared by dissolving 0.28 g of lime CaO in enough water to make 1.00 L of limewater $[Ca(OH)_2 (aq)]$.

2.10.8. What operations would you follow to prepare exactly 100 mL of 0.200 M acetate buffer, pH 5.00, starting with pure liquid acetic acid and solutions containing 3 M HCl and 3 M NaOH?

2.10.9. Look up pKa for each of the following acids and decide which one would be best for preparing a buffer of pH 3.10
    (i) hydroxybenzene pKa = 9.98
    (ii) propanoic acid pKa = 4.87
    (iii) cyanoacetic acid 2.47
    (iv) sulfuric acid 1.99 sulphuric if using Br Eng

2.10.10. From the K_b values listed below, which of the following bases would be best for preparing a buffer of pH 9.0?
    (i) NH_3 (ammonia, $K_b = 1.8 \times 10^{-5}$)

(ii) $C_6H_5NH_2$ (aniline, $K_b = 4.0 \ x \ 10^{-10}$)

(iii) H_2NNH_2 (hydrazine, $K_b = 3.0 \ x \ 10^{-6}$)

(iv) C_5H_5N (pyridine, $1.6 \ x \ 10^{-9}$)

2.10.11. Why does buffer capacity increase as the concentration of buffer increases?

2.10.12. Calculate how many milliliters of 0.100 M HCl should be added to how many grams of sodium acetate dihydrate ($NaOAc.2H_2O$, FM 118.06) to prepare 250.0 mL of 0.100 M buffer, pH 5.00.

2.11 References

Barrett, R. L. (1955). The Formaldehyde Clock Reaction, *J. Chem. Ed., 32,* 78.

Baumgarner. R. E., Lavert, Jr., T. F., Rogers, C. M. and Isil, S. S. (2002). Cloudwater Chemistry from Ten Sites in North America, *Environ. Sci. Technol. 36,* 2614.

D'Alba, I., Carloni, I. and Ferrate, A. L. (2014). Hyperventilation Syndrom in Adolescents with and Without Asthma, *Padiatr. Pulmonol, 10,* 23145.

de Levie, R. (2003). The Henderson-Hasselbalch Equation: Its History and Limitation, *J. Chem. Ed., 80,* 146.

Griffin, C., Rosenzweig, J. Ji, N. and Rosenzweig, Z. (2000). Synthesis and Application of Submicrometer Fluorescence Sensing Particles for Lysosomal pH Measurements in Marine Macrophages, *Anal. Chem. 72,* 3497.

Gross, L. M. (2003). A Demonstration of Acid Rain and Lake Acidification: Wet Deposition of Sulfur, *J. Chem. Ed., 80,* 39.

Harris, D. C. (2005). *Exploring Chemical Analysis,* 3rd ed. W. H. Freeman and Company, New York NJ.

John, S. T. (1977). The Pathway to the Ostwald Dilution Law, *J. Chem. Ed. 74,* 865.

Kang, S. B. (2015). A Case of Hyperventilation Syndrom Mimicking Partial Seizure: Usefulness of EEG Monitoring in Emergency Department, *J. Epilepsy Res. 5,* 20.

Kern, M. (1960). The Hydrationof Carbon dioxide, *J. Chem. Ed. 37,* 14.

Lower, S. K. (2001). Chemical Equilibrium, Original Frame Maker, Canada.

Riley, J. T. (1977). Display of Sodium as a Shiny Metal, *J. Chem. Ed., 54,* 29.

White, M. A. (1998). The Chemistry Behind Carbonless Copy Paper, *J. Chem. Ed., 75,* 1119.

Chapter 3

SOLUBILITY AND SOLUBILITY PRODUCT

3.1   Solubility

Solubility (S) is the maximum amount of substance that dissolves completely in a given amount of solvent at a given temperature to produce a stable solution. This is commonly measured in $g.dm^{-3}$, $mole/dm^3$ or $g/100$ g. The objective here is to be more explicit when:

as much solute is dissolved as can be dissolved at room temperature, the solution is *saturated*.

less solute is dissolved than can be dissolved at room temperature, the solution is *unsaturated*, and

more solute is dissolved than can be dissolved at room temperature, the solution is *supersaturated*.

Solids show a wide range of solubility in water, for example Table 3-1 below:

Table 3.1: Ionic and Covalent Compounds Solubilities in Water

Types of substance	Substance	Solubility at 298 K $g.dm^{-3}$
Ionic	KCl	4.2×10^2
	NaCl	3.6×10^2
	$PbCl_2$	1.1×10^1
	$BaSO_4$	2.3×10^{-3}
Covalent	Glucose	2.0×10^3
	Bromine	3.2×10^1

Some solids may or may not dissolve depending on the nature of the force acting between solute and solvent particles. Usually, the solubility of solids increases with increasing temperature. Unlike solids, the solubility of nearly all gases decreases with an increase in temperature, showing that enthalpy of solution is exothermic. Think of an aqueous solution of a slightly soluble substance of type MA. The following equilibrium expression exists in a saturated solution:

Under the circumstances where there is no other species of M and A is present in the solution, the molar solubility of MA is given by:

$$S = [MA] + [M^+] = [MA] + [A^-] \qquad (3.1)$$

[MA] is the molecular solubility of MA, which is often negligibly small compared to [M$^+$] and [A$^-$] for compounds that are highly ionized. Even salts like silver chloride and calcium sulfate have determinable molecular solubilities; the undissociated species of the latter is interpreted as an ion pair (Ag$^+$ Cl$^-$). It has been mentioned previously that **S** is a function of temperature. Its actual value depends on the nature of crystal of MA (for example, the solubility of calcite $CaCO_3$ rhombohedral, and is not the same as that of aragonite, orthorhombic $CaCO_3$) and to some certain extent the particle size. The previous fact was pointed out by Wollaston as early as 1813, who found that finely divided solids have a greater solubility than larger crystals.

The **Law of Mass** action shows the concentrations equal to activities in (3.2)

$$K = \frac{[M^+][A^-]}{[MA]}$$

(3.2)

where K is the ionization constant of MA. In a saturated solution at a specified temperature if you rearrange expression 3.2 you will obtain the following expression 3.3

$$[M^+][A^-] = K\,[MA]_{salt} = constant = Ksp$$

(3.3)

That constant is known as the solubility product constant of MA. For a precipitate M_xA_y,

$$[M^{m+}]^x[A^{a-}]^y = Ksp$$

(3.4)

Once the solubility of the precipitate is clearly known and is so small compared to the concentration of the dissociated species that it can be neglected, and no other form of M and A (such as complex species MA_2^-, etc.) are present in the saturated solution, the molar solubility of MA can be calculated as in expression 3.5 below:

$$S = [M^+] = [A^-] = (Ksp)^{1/2}$$

(3.5)

It is important to note the role of the *stoichiometry* of the dissolving process in these calculations. The stoichiometry of the dissolving process, which shows the number of ions of individual type formed by the dissolution process and therefore the powers to which the concentrations of the ions must be raised, also tells the relative concentrations of the ions involved. Thus, the dissolution of x mol/L of MgF_2 produces x mol/L of Mg^{2+} and $2x$ mol/L of F^-; therefore, the expression for calculating molar solubility of MgF_2 (s) in water at 25 ºC is given by x, is $[Mg^{2+}][F^-]^2 = (x)(2x)^2 = Ksp$.

The expression used to calculate the molar solubility of a salt depends on the stoichiometry of the dissolving process; thus, we can use Ksp values of salts like stoichiometry for predicting relative solubilities. For example, one can use the decreasing Ksp values for AgCl, AgBr, and AgI to predict their relative solubilities, which are in the following trend $S(AgCl) < S(AgBr) < S(AgI)$, but it is not possible to compare the Ksp value of AgI to the Ksp value of PbI_2 to predict which of these salts will be more soluble. The reason being that the molar solubility x of AgI is related to its Ksp by $(x)(x) = Ksp$, while that of PbI_2 is related to its Ksp by $(x)(2x)^2 = Ksp$. The solubility of a substance can be used to work out the solubility product.

Example 3.1

The solubility of $BaSO_4$ is 2.33×10^{-3} g.dm^{-3} at room temperature; FM of $BaSO_4$ is 233 g.mol^{-1}.

$$BaSO_4\,(aq) \rightleftharpoons Ba^{2+}(aq) \;+\; SO_4^{2-}\,(aq)$$

Intial $\quad \dfrac{2.33 \times 10^{-3}}{233} \qquad\qquad — \qquad\qquad — \qquad$ mole. dm^{-3}

At equilibrium $\;—\; \qquad \dfrac{2.33 \times 10^{-3}}{233} \qquad \dfrac{2.33 \times 10^{-3}}{233} \qquad$ mole. dm^{-3}

$Ksp = [Ba^{2+}]\,[SO_4^{2-}] = [10^{-5}][10^{-5}]$ mol^2.dm^{-6}

$Ksp = 1 \times 10^{-10}$ mol^2. dm^{-6}

Example 3.2

Given that Ksp for iron (III) hydroxide at 298 K is 7.9×10^{-16} mol^3. dm^{-9} and its FM is 90 g. mol^{-1}. Calculate its solubility in pure water.

Solution

$$(3.6)$$

$$Ksp = [Fe^{2+}][OH]^2 \tag{3.7}$$

Let the solubility be mol.dm^{-3} at 298 K, and at equilibrium you have the following:

$$(3.8)$$

Upon substituting expression 3.8 into expression 3.7, the following expression is obtained as

$(x).(2x)^2$ (mol. dm^{-3}).(mol. dm^{-3})^2 = 4 x^3 mol.^{3} dm^{-9} = 7.9 x 10^{-16} mol.^{3} dm^{-9}
x = 5.82 x 10^{-6} mol. dm^{-3}

Therefore, the solubility of Fe(OH)$_2$ in water is
5.82 x 10^{-6} mol. dm^{-3} x 90 g. mol^{-1} = 5.24 x 10^{-4} g. dm^{-3}

Practice 3.1
Which has the greater molar solubility: AgCl with Ksp = 1.8 x 10^{-10} or
Ag$_2$CrO$_4$ with Ksp = 1.1 x 10^{-12}? Which has the greater solubility in grams per
litre? **Solution:** AgCl solubility = 1.3 x 10^{-5} M (0.0019 g/L), Ag$_2$CrO$_4$
solubility = 6.5 x 10^{-5} M (0.0216 g/L) has both the higher molar and gram
solubility, despite its smaller value of Ksp.

3.2   Solubility Product

Compounds having a solubility of less than 0.01 mol of dissolved material
per litre of solution are often considered to be insoluble. Most compounds
that are classified as being insoluble in water dissolve to a certain extent
anyway. The extent to which an insoluble compound dissolves in water at
25 ^{o}C can be expressed in terms of the equilibrium constant for the
dissolving process. That constant is called the solubility product constant
and it is designated Ksp. The solubility product, Ksp, is the equilibrium
constant for the reaction in which a solid salt (an ionic compound) dissolves
to give its constituent ions in solution. The concentration of the solid is
omitted from the equilibrium constant because the solid is in its standard
state: see expression 3.3.

$$K = \frac{[Ag^+][Cl^-]}{AgCl}$$

AgCl being constant will change the above expression to become
$$Ksp = [Ag^+][Cl^-]$$
Ksp is the solubility product of the salt.

Definition
Solubility product of the sparingly soluble binary electrolyte is the product of
the concentration of the ions in a saturated solution raised to the appropriate
powers, at a given temperature as indicated in expression 3.4.

General A$_x$B$_y$ $\rightleftharpoons$ xA$^+$ + yB$^-$

[M$^+$]^x[A$^-$]^y = Ksp

The main function of solubility product deals with the conditions under which an electrolyte will dissolve or come out of the solution as a solid forming precipitate.

The following are the possibilities:

(1) a solution in which [M$^+$] [A$^-$] is less than Ksp of MA is not saturated and more of MA can be dissolved.

(2) a solution in which [M$^+$] [A$^-$] is equal to the Ksp (MA) the solution is saturated, and

(3) if anything to be done to a solution, which tends to make the [M$^+$] [A$^-$] to be greater than Ksp of the solid MA will be precipitated. Evaporation of a saturated solution tends to increase ionic concentration of the ions present. As a result, the solid is deposited.

Example 3.3

The following is an example of how to calculate the solubility of an ionic compound.

Consider the dissociation of Lead (II) iodide in water, Ksp for PbI$_2$ is 7.9 x 10^{-9}.

$$Ksp = 1.9 \text{ x } 10^{-9} = [Pb^{2+}][I^-]^2$$

A solution that contains the entire solid capable of being dissolved is said to be saturated. What is the concentration of Pb^{2+} in a solution saturated with PbI$_2$?

Solution:

Looking at expression 3.3, two I$^-$ (2 I$^-$) ions are produced for each Pb^{2+} ion. If the concentration of the dissolved Pb^{2+} is x M, the concentration of dissolved I$^-$ must be $2x$ M This can be displayed as follows:

	PbI$_2$ (s) $\rightleftharpoons$	Pb^{2+}(aq) +	2 I$^-$ (aq)
Initial concentration	solid	0	0
Final concentration	solid	x	$2x$

Substituting these concentrations into the solubility product reaction equation (3.3) gives

$$[Pb^{2+}][I^-]^2 = (x)(2x)^2 = 7.9 \times 10^{-9}$$
$$4x^3 = 7.9 \times 10^{-9}$$
$$x = \sqrt[3]{\frac{7.9 \times 10^{-9}}{4}}$$
$$x = 0.0012_5$$

The concentration of Pb^{2+} is 0.0012_5 M and the concentration of I^- is $2x = (2)(0.0012_5)$

$2x = 0.0025$ M. The physical meaning of the solubility product is this: if an aqueous solution is left in contact with excess solid of PbI_2, the solid will dissolve until the condition $[Pb^{2+}][I^-]^2 = K_{sp}$ is satisfied. Thereafter, no more solid dissolves. Unless excess solid remains, there is no guarantee that $[Pb^{2+}][I^-]^2 = K_{sp}$. If the product of $[Pb^{2+}][I^-]^2$ exceeds the K_{sp}, then PbI_2 (s) will precipitate.

Practice 3.2.

The solubility product of silver chromate (VI) is 2.5×10^{-12} mol^2.dm^{-9}. What will be its solubility? **Solution** is 8.55×10^{-5} mol.dm^{-3}.

There is substantial difference between the *solubility* of a salt or compound to the solubility product of that salt. The solubility of a salt usually refers to the maximum quantity of salt that can be dissolved in a given volume; the quantity of salt that in fact is dissolved in a given volume of a *saturated solution* is expressed in grams per 100 mL, mole per litre, or any other units. Furthermore, the *solubility product* is an equilibrium constant. However, there is a concentration between these two quantities: once one of them is known, the other can be easily calculated.

3.3   Formation of Precipitate

The goal of this part is to develop a method for determining whether an insoluble salt will form a precipitate under a given set of concentrations of its constituent ions. It is imperative for a chemist to know with confidence if a precipitate will form when two solutions are mixed. The mixture which is not yet at equilibrium is described by the reaction quotient, Q. For example, a solution that contains calcium ions is mixed with a solution that contains phosphate ions, this can be expressed as:

Though the reaction gives calcium phosphate, for dissolution of the solid, the reverse of the reaction is the way that the value for the equilibrium is tabulated. Thus, the reaction quotient is calculated as:

$Q = [Ca^{2+}]_t^3[PO_4^{3-}]_t^2$ and Q is compared with K_{sp} of the compound to determine the direction. Here, Q is called the *ionic product* for apparent reasons. The expression for ionic product is initial concentration that has arbitrary concentrations at time t, not necessarily equilibrium concentration. This can clearly be expressed by a number line of solubility, can be seen below:

Before analyzing the solubility number line it is important to bear in mind that Q, uses *initial concentrations* rather than *equilibrium concentrations,* which is the case for *Ksp.* The analysis of the solubility number line can be interpreted as follows:

Q1 is smaller than *Ksp;* when Q is less than *Ksp* a reaction will occur towards the right and some additional solid will dissolve, forming more ions. Q2 is greater than *Ksp* and additional precipitation will occur under these conditions; as a result the reaction occurs towards the left.

A solution in which Q is less than *Ksp* is unsaturated and the system is not at equilibrium: this is a solution in which more material can be dissolved. When Q is equal to *Ksp,* the solution is saturated and no additional solute can be dissolved. As can be seen from the number line, Q2 represents the condition of a *supersaturated* solution, since Q exceeds *Ksp.* Therefore, *supersaturation* is that condition at which a solution contains more of the solute in solution than it could hold at that temperature if crystals of solute were present. In other words, the system is not at equilibrium.

3.3.1 Ksp and Precipitation Reactions

We can use a comparison of Q against *Ksp* to determine the following: (1) if a precipitate will form when the concentrations of the ions are known or can be calculated, and (2) what concentration of an ion is required to begin the precipitation of an insoluble salt when the concentration of the other ion is known.

Example 3.4

How to determine whether a precipitate will occur or not: Suppose 500 cm^3 of 0.02 M NaCl are added to 500 cm^3 of 0.01M AgNO$_3$. If *Ksp* of AgCl is 1.78 x 10^{-10}, will a precipitate of AgCl form?

Solution

We have to determine Q and compare it to its *Ksp*

Total volume of solution will be 1000 cm³, because 500 cm³ + 500 cm³ of each solution are used. Therefore, the concentrations of both the NaCl and Ag NO₃ solution will be halved.

$$[NaCl] = \frac{1 \; x \; 0.02 \; M}{2} = 0.01 \; M$$

$$[AgCl] = \frac{1 \; x \; 0.01 \; M}{2} = 0.005 \; M$$

Assuming both salts dissociate to completion, therefore:

Q = [Ag⁺][Cl⁻]
   = (0.005) (0.01)
Q = 5.00 x 10⁻⁵

The numerical value for *Ksp* is 1.78 x 10⁻¹⁰. Since Q, ionic product is 5.0 x 10⁻⁵ this exceeds *Ksp*, therefore AgCl will precipitate out of solution until [Ag⁺][Cl⁻] = 1.78 x 10⁻¹⁰

3.3.2 Separation of Ions through Selective Precipitation

It is always easy to separate a mixture of ions by adding a solution that will precipitate some ions and leave others behind. Consider a situation where you have a mixture of Cl⁻ and SO₄²⁻ in one beaker, for example, which can be separated by addition of Ba(NO₃)₂. This will result in the formation of Ba(SO₄)₂ precipitates, but Cl⁻ ions will remain in solution because the BaCl₂ is soluble. On the other hand, Ag⁺ and Zn²⁺ cations can be separated by the addition of dilute HCl. Silver chloride, AgCl will precipitate out leaving Zn²⁺ in solution because ZnCl₂ is soluble.

3.4   Common Ion Effect

What is common ion effect?

This is a decrease in solubility due to the presence of a solution that contains an ion in common with the precipitate or is the effect due to decrease in solubility of sparingly soluble substance resulting from the addition of a solution that provides one of the ions, which was already present in the saturated solution of sparing soluble substance. This has something to do with the application of Le Chatelier's principle. Consider the following;

This will be affected by the addition of either more M⁺ or more A⁻ ions. The expression of the solubility of a sparingly soluble ionic substance by addition of an excess of one of its ions is important in gravimetric determinations and separations by precipitation in some titrations.

The source of common ion can be a strong acid (a source of H_3O^+), strong base (a source of OH^-) or a soluble ionic salt, including a soluble ionic salt that is formed during the reaction of a weak acid or weak base. The common ion is often a product which opposes the ionization of the weak acid or base in accordance to Le Chatelier's principle.

What happens when we add a second source of I^- to a solution saturated with PbI_2 (s)?

Example 3.5
Suppose you are adding 0.30 M NaI, which dissociates completely to Na^+ and I^-. What is the concentration of Pb^{2+} in this solution?

The initial concentration of I^- is from dissolved NaI. The final concentration of I^- has contributions from NaI and PbI_2.

$$Ksp \implies [Pb^{2+}][I^-]^2 = (x)(2x + 0.030) = 7.9 \times 10^{-9} \tag{3.10}$$

Solution
The value of x, without additional I^- was found to be $x = 0.0012_5$ M. In the present case we anticipate that $x \ll$ than 0.0.00125 M, because of Le Chatelier's principle. Addition of I^- in expression 3.9 displaces the reaction in the reverse direction. We suspect that in expression 3.10, $2x$ may be smaller than 0.030. As an approximation we can ignore $2x$ in comparison with 0.030. Thus, expression 3.10 simplifies to $(x)(0.030)^2 = Ksp = 7.9 \times 10^{-9}$. Because $2x = 1.8 \times 10^{-5} \ll 0.030$, it is possible to ignore $2x$ in solving this problem. In the absence of added I^- the solubility of Pb^{2+} was 0.0012_5 M. On the other hand, the presence of 0.030 M I^- causes the concentration of Pb^{2+} to be reduced to 8.8×10^{-6} M. It should be clearly noted that the concentration of the common ion added is much larger than the concentration of that ion coming from the insoluble salt, so the concentration of ion coming from the insoluble salts can be mathematically neglected with respect to the concentration of the common ion. This is almost often the case, but it is advisable to check the approximation before neglecting the concentrations contributed by insoluble salt ions.
Practice 3.3
Calculate the solubility of manganese (II) sulphide.

(a) in water and (b) in a 1.0 x 10^{-10} mol. dm^{-3} solution of suphide ion. Ksp (MnS) at 25 ^{o}C = 2.5 x 10^{-13} mol^2. dm^{-6} and compare the results of (a) and (b). **Solution** (a) 5.0 x 10^{-7} mol.dm^{-3} (b) Mn^{2+} = 2.5 x 10^{-11} mol. dm^{-3}.

3.5   Application of Solubility Product in Qualitative Analysis
This can be briefly explained using the following example:

Group II
Precipitation of sulphides of Hg^{2+}, Cu^{2+}, Cd^{2+}, Sn^{2+}, Sn^{4+} by use of H_2S gas or $(NH_4)_2S$ solution in acidic condition of dilute HCl.

$$Ksp = [M^{2+}][S^{2-}]$$

The question is, what is the role of HCl in the above precipitation?

$$MS \rightleftharpoons M^{2+} + S^{2-} \tag{3.11}$$

$$H_2S \rightleftharpoons H^+ + SH^- \rightleftharpoons 2H^+ + S^{2-} \tag{3.12}$$

$$HCl \rightleftharpoons H^+ + Cl^-$$

The dilute HCl increases the H^+ ions in solution; hence decreasing the concentration of S^{2-} ions in expressions 3.11 and 3.12, which hinders the precipitates of sulphides. Therefore, sulphides with small solubility product are precipitated. In addition, when a weak acid is used, the precipitates of sulphides with small solubility product result in very high $[S^{2-}]$. The latter will cause even CoS, which does not belong to group II: see Table 3-1 to precipitate as well.

Table 3.2: Group II and IV Compounds Ksp

Group II	Ksp	Group IV	Ksp
HgS	3 x 10^{-54}	CoS	3 x 10^{-26}
CuS	3 x 10^{-42}	ZnS	1 x 10^{-23}
CdS	4 x 10^{-29}	NiS	4 x 10^{-21}
PbS	4 x 10^{-28}	MnS	1.5 x 10^{-15}

Group III
The precipitation of hydroxides compounds of Fe^{3+}, Al^{3+}, and Cr^{3+}. In order to limit precipitation of metal hydroxides sufficient OH^- ions must be added to exceed their solubility product, but it should not be higher than those in Table 3-2.

Compound	Ksp
$Fe(OH)_3$	1×10^{-38}
$Al(OH)_3$	1×10^{-23}
$Cr(OH)_3$	1×10^{-30}

How products are precipitated in presence of NH_4Cl salt.

$$NH_3\,(g) \quad + \quad H_2O\,(l) \quad \rightleftharpoons \quad NH_4^+\,(aq) \quad + \quad OH^-\,(aq)$$

The presence of NH_4Cl increases the $[NH_4^+]$, which suppresses the ionization of ammonium solution, in this way limiting high $[OH^-]$. The concentration of hydroxyl ions is therefore sufficiently large enough to precipitate the hydroxides of iron, aluminium, and chromium, but insufficient to precipitate the hydroxides of the other metals. In order to precipitate iron (III) in group (II) we must convert Fe (II) to Fe (III) (oxidation process) by boiling the solution in group (II) by just the addition of one or two drops of concentrated HNO_3.

3.6 Simultaneous Equilibria

In many cases two or more equilibrium processes occur at the same time in solution. Chemists often consider such cases as examples of simultaneous equilibria. A good example of simultaneous equilibria involves the addition of a reagent to a saturated solution of an insoluble salt converting the salt to a second less soluble salt. Think of a situation when you add a few drops of K_2CrO_4 to a saturated solution of $PbCl_2$ converting the white $PbCl_2$ to yellow $PbCrO_4$. The reaction that has taken place can be considered as the sum of two other reactions (3.13 and 3.14), namely

Dissolving of PbCl₂:

$$PbCl_2\,(s) \;\rightleftharpoons\; Pb^{2+}\,(aq) + 2\,Cl^-\,(aq) \qquad K_1 = Ksp = 1.7 \times 10^{-6} \qquad (3.13)$$

Precipitation of PbCrO₄:

$$K_2 = \frac{1}{K_{sp}} = \frac{1}{1.8 \times 10^{-14}} \qquad (3.14)$$

Net reaction:

$$K_{net} = K_1 . K_2 = 9.4 \times 10^8$$

Therefore, the large value for K_{net} indicates the reaction should be product - favoured.

3.7 Solubility and pH of the Solution

A basic anionic compound becomes more soluble as the acidity of the solution increases. A good example is the solubility of $CaCO_3$, which increases with the decrease in pH because the CO_3^{2-} ions combine with protons to form HCO_3^- ions. Upon removal of CO_3^{2-} ions from the solution, the solubility equilibrium will shift to remove that constraint as predicted by Le Chatelier's principle in equation 3.16. The overall reaction is the dissolution of $CaCO_3$ in acidic solution to produce Ca^{2+} ions and HCO_3^- ions. Also, the solubility of salts with such basic anions as CN^-, PO_4^{3-}, S^{2-} or F^- increases as the acidity of the solution increases.

$$\left.\begin{array}{l} CaCO_3\,(s) \rightleftharpoons Ca^{2+}\,(aq) + CO_3^{2-}\,(aq) \\[6pt] \underline{H_3O^+\,(aq) + CO_3^{2-}\,_{(aq)} \rightleftharpoons HCO_3^-\,(aq) + H_2O\,(l)} \\[6pt] CaCO_3\,(s) + H_3O^+\,(aq) \rightleftharpoons Ca^{2+}\,(aq) + H_2O\,(l) \end{array}\right\} \quad (3.16)$$

The metal hydroxide solubility is driven by the union of hydronium ion from the acids and hydroxide ions from the equilibrium of the insoluble hydroxide. For example, dissolution of magnesium hydroxide in acid solution can take place in the following way:

$$Mg(OH)_2\,(s) \rightleftharpoons Mg^{2+}\,(aq) + 2\,OH^- \qquad Ksp = 1.5 \times 10^{-11} \qquad (3.17)$$

Union of hydronium ion and hydroxide ion:

$$K = \frac{1}{K_w} = 1.0 \times 10^{14} \qquad (3.18)$$

Net reaction:

$$Mg(OH)_2\,(s) + 2\,H_3O^+\,(aq) \rightleftharpoons Mg^{2+}\,(aq) + H_2O\,(l) \qquad (3.19)$$

$$K_{net} = Ksp.K^2 = 1.5 \times 10^{17}$$

where K is used to calculate K_{net}, because it is necessary to multiply 19 g by two $(2\,H_3O^+)$ to obtain net reaction.

Example 3.6

What concentration of OH^- must be exceeded in a 0.010 M solution of $Ni(NO_3)_2$ in order to precipitate $Ni(OH)_2$? (Assume that the added OH^- does not change the concentration of Ni^{2+}).

Solution:

We know that $Ksp = 1.6 \times 10^{-14}$ for $Ni(OH)_2$. In case of any excess OH^- in a saturated solution of $Ni(OH)_2$ will cause some $Ni(OH)_2$ to precipitate. For saturated solution we have:

$Ksp = [Ni^{2+}][OH^-]^2 = 1.6 \times 10^{-14}$

Therefore, if $Q = [Ni^{2+}][OH^-]^2 > 1.6 \times 10^{-14}$, precipitation will occur. Suppose $[OH^-] = x$ and using $[Ni^{2+}] = 0.010$ M, we have

$(0.0010)\,x^2 > 1.6 \times 10^{-14}$

$$x^2 > \frac{1.6 \times 10^{-14}}{0.010} = 1.6 \times 10^{-12}$$

$$x > \sqrt{1.6 \times 10^{-12}} = 1.3 \times 10^{-6} \; M$$

This means $Ni(HO)_2$ will precipitate if OH^- exceeds 1.3×10^{-6} M. The latter concentration corresponds to a solution pH of 8.11 (pH = 14.00 – pOH = 14.00 - 5.89 = 8.11). Thus, $Ni(HO)_2$ will precipitate when the solution pH is 8.11 or higher.

Conversely, pH has no effect on the solubility of compounds that contain anions of strong acids (Cl^-, Br^-, I^-, NO_3^- and ClO_4^-), the reason being that these anions cannot be protonated by H_3O^+. In addition, the effect of pH on solubility plays a great role in understanding how fluoride iron reduces tooth decay. At the time when tooth enamel comes into contact with F^- ions in fluorine containing toothpaste or drinking water, OH^- ions in hydroxyapatite ($Ca_5(PO_4)_3OH$) are replaced by F^- ions, giving the mineral fluoroapatite ($Ca_5(PO_4)_3F$). The F^- being a weaker base than OH^-, $Ca_5(PO_4)_3F$ is much more resistant than $Ca_5(PO_4)_3OH$ to dissolve in acid.

Practice 3.4
From the following compounds list which are more soluble in acidic solution than in water (a) ZnS (b) $Al(OH)_3$ (c) PbI_2 (d) AgCN **solution** AgCN, $Al(OH)_3$, and ZnS.

Metal cations, M^{2+}, can be separated into two groups through selective precipitation of metal sulfides, MS. A good example is Pb^{2+}, Cu^{2+}, and Hg^{2+}, which form insoluble sulfides, and are often separated from Mn^{2+}, Fe^{2+}, Co^{2+}, Ni^{2+} and Zn^{2+}, which form more soluble sulfides. The separation carried out under acidic condition utilizes the following solubility equilibrium as indicated in expression 3.20.

$$MS \; (s) \; + \; 2\,H_3O^+ \,(aq) \; \rightleftharpoons \; M^{2+}\,(aq) \; + \; H_2S \,(aq) \; + \; 2\,H_2O \,(l) \qquad (3.20)$$

The equilibrium constant derived from expression 3.20 is called solubility product in acid, and gives the following symbol K_{spa} in expression 3.21.

$$K_{spa} = \frac{[M^{2+}][H_2 S]}{[H_3O^+]^2}$$

$$(3.21)$$

The adjustment of H_3O^+ concentration controls the separation such that the reaction quotient Q_c exceeds K_{spa} for insoluble sulfides, but not for the more soluble ones, as can be seen in Table 3.3.

Table 3.4: Solubility Product in Acid (K_{spa}) at 25 °C for Metal Sulfides

Metal Sulfides, MS	K_{spa}	Metal Sulfide, MS	K_{spa}
MnS	3×10^{10}	ZnS	3×10^{-2}
FeS	6×10^{2}	PbS	3×10^{-7}
CoS	3	CuS	6×10^{-16}
NiS	8×10^{-1}	HgS	2×10^{-32}

Solubility product for acid (*spa*) is used instead of solubility product for several reasons, one being that the ion separations are carried out in acidic solution, therefore it makes sense to use *spa*. In addition, most Ksp values found in the literature for the general reaction 3.22 below:

$$\text{MS (s)} \quad \rightleftharpoons \quad M^{2+}\text{(aq)} \; + \; S^{2-}\text{(aq)} \tag{3.22}$$

are not quite correct because they are based on K_{a2} value for H_2S (1.3×10^{-14}) which has a significant error. In actual fact, the correct value of K_{a2} for H_2S (10^{-19}) is relatively small, which means S^{2-} is basic and has less significance in aqueous solutions. Hydrogen sulfide remains as a major sulfide containing specie in aqueous acidic solutions and HS^- in basic solutions. The following in expression 3.23 shows the existing general relationship between K_{spa} and K_{sp}, K_{a1} and K_{a2} values.

$$K_{spa} = \frac{K_{sp}}{K_{a1} K_{a2}} \tag{3.23}$$

Suppose in a given experiment you have M^{2+} concentrations which are approximately 0.01 M, and the H_3O^+ concentration is about 0.30 M by adding HCl. This solution is saturated with H_2S gas, which makes an H_2S concentration become approximately 0.10 M. When these concentrations are substituted into the equilibrium constant expression, the reaction quotient Q_c is found to be 1×10^{-2}.

$$Q_c = \frac{\left[M^{2+}\right][H_2 S]}{[H_3O^+]^2} = \frac{(0.01)(0.10)}{(0.30)^2} = 1 \times 10^{-2}$$

Obviously, the above value of Q_c exceeds Ksp for PbS, CuS, and HgS (Table 3.3), but does not exceed K_{spa} for MnS, FeS, CoS, NiS or ZnS. Consequently, PbS, CuS, and HgS precipitate under these acidic conditions, but Mn^{2+}, Fe^{2+}, Co^{2+}, Ni^{2+}, and Zn^{2+} remain in solution.

3.8   Solubility and Complex Ions

It is apparent that the solubility of insoluble salts can be improved dramatically by the formation of complex ions of the metal ion in Lewis-base reactions, which form a coordinate covalent bond to the metal cation. In short, a *complex* ion is one which contains a metal cation bonded to one or more small molecules or ions such as CN^-, OH^- or NH_3. For example, AgCl readily dissolves in NH_3 solution due to the formation of $Ag(NH_3)_2^+$ (see expressions 3.24, 3.25, and 3.26). This shows how ammonia is tying up Ag^+ and shifting solubility equilibrium in accordance with Le Chatelier's principle.

Dissolving of AgCl:

$$AgCl\,(s) \rightleftharpoons Ag^+\,(aq) \quad + \quad Cl^-\,(aq) \qquad Ksp = 1.8\ x\ 10^{-10} \qquad (3.24)$$

Complexing of Ag^+:

$$Ag^+\,(aq) \quad + \quad 2\,NH_3\,(aq) \rightleftharpoons Ag(NH_3)_2^+\,(aq) \quad K = 1.6\ x\ 10^7 \qquad (3.25)$$

Net reaction:

$$AgCl\,(aq) \ + \ 2\,NH_3\,(aq) \rightleftharpoons Ag(NH_3)_2^+\,(aq) \qquad K_{net} = K_{sp}.K \qquad (3.26)$$
$$= 2.9\ x\ 10^{-3}$$

Indeed, the above series of reactions show the sequential formation of insoluble silver salts, and soluble silver ion complex expression 3.26 is of commercial value, because it is the reaction that is used *to remove unexposed AgBr from black and white photographic film* to avoid having entire film exposed during the developing process. Furthermore, the formation of complex ions is a stepwise process involving many species, so the equilibrium expression is quite complicated. The situation becomes simple if the ligand is present in excess: see expressions 3.27 and 3.28. For example, in the formation of $CuCl_4^{2-}$, the equilibrium is as follows:

$$[Cu(H_2O)_6]^{2+}\,(s) \ + \ 4\,Cl^-\,(aq) \rightleftharpoons CuCl_4^{2-}\,(aq) \ + \ 6\,H_2O\,(l) \qquad (3.27)$$

The concentration of water is constant, therefore:
$$K_{stab.} = \frac{[CuCl_4^{2-}]}{[Cu(OH)_6]_2\,[Cu^-]_4}$$

$$K_{stab.} = \frac{[CuCl_4^{2-}]}{[Cu(OH)_6]_2\,[Cu^-]_4} \qquad (3.28)$$

whereas $K_{stab.}$ is stability constant: the larger the value of this constant the more stable the complex ions. When the ligand is displaced from metal ion by another ligand, this can lead to the formation of a complex with a larger stability constant, e.g.,

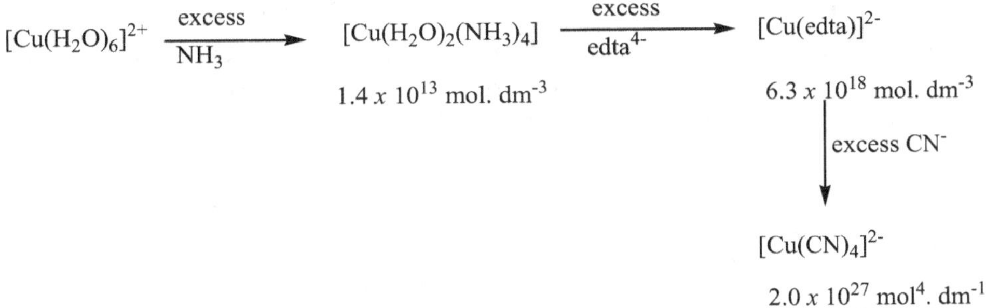

3.9 Problems

3.9.1. (a) What is common-ion effect? (b) Provide an example of a salt that can decrease the ionization of HOCl in solution. (c) Explain why ionization of a weak base is suppressed by the presence of its conjugate acid.

3.9.2. Assume that 0.010 M NaF solution is slowly added, with stirring, to a solution that contains 0.10 M $Cr(NO_3)_3$ and 0.10 M $Ca(NO_3)_2$. Which salt-CrF_3 or CaF_2 will precipitate first? (Ksp for CaF_2 is 5.3 x 10^{-9}; for CrF_3 it is 6.6 x 10^{-11}).

3.9.3. Will precipitate form when 0.10 L of 3.0 x 10^{-3} M $Pb(NO_3)_2$ is added to 0.40 L of 5.0 x 10^{-3} M Na_2SO_4?

3.9.4. The solubility product for gadolinium hydroxide, $Gd(OH)_3$, is 1.8 x 10^{-23}. If a solution is 0.010 M in Gd^{3+} ion and the pH of the solution is slowly increased, at what pH will $Gd(OH)_3$ begin to precipitate?

3.9.5. Describe the effect of pH (increase, decrease, or no change) that results from the following additions: (a) Sodium formate, $NaCHO_2$, to a solution of formic acid, $HCHO_2$ (b) Ammonium perchlorite, NH_4ClO_4, to a solution of ammonia, NH_3, (c) Potassium bromide to a solution of potassium nitrate, KNO_3, (d) Hydrochloric acid, HCl, to a solution of sodium acetate, $NaC_2H_3O_2$.

3.9.6. A 20.0 mL sample of 0.20 M HCl solution is titrated with 0.20 M NaOH solution. Calculate the pH of the solution after the following volumes of base have been added: (a) 15.0 mL (b) 20 mL (c) 19.9 mL

3.9.7. Calculate [OH-] and pH for each of following strong base solutions: (a) 2.33 g of NaOH in 500 mL of solution (b) 10.0 mL of 0.150 M calcium hydroxide diluted to 500 mL.

3.9.8. A 50.00 mL solution containing NaBr was treated with excess $AgNO_3$ to precipitate 0.214 g of AgBr (FM 187.772).
(a) What number of moles of AgBr products were isolated?
(b) What was the molarity of NaBr in the solution?

3.9.9. Calculate the pH of a solution prepared by dissolving 0.28 g of lime (CaO) in enough water to make 1.0 L limewater [$Ca(OH)_2$ aq].

3.9.10. Prior to having an X-ray examination of the upper gastrointestestinal tract, a patient drinks an aqueous suspension of solid $BaSO_4$. (Scattering of X-rays by barium greatly enhances the quality of the photograph). Although Ba^{2+} is toxic, ingestion of $BaSO_4$ is safe because it is quite insoluble. If a saturated solution prepared by dissolving solid $BaSO_4$ in water has $[Ba^{2+}] = 1.05 \, x \, 10^{-5}$ M, what is the value of Ksp for $BaSO_4$?

3.9.11. Determine whether Cd^{2+}, can be separated from Zn^{2+} by bubbling H_2S through a 0.30 M HCl that contains 0.005 M Cd^{2+} and 0.005 M Zn^{2+}, (K_{spa} for CdS and ZnS are $8 \, x \, 10^{-7}$, $3 \, x \, 10^{-2}$ respectively).

3.9.12. Use the Henderson-Hasselbalch equation to calculate the pH of a buffer solution that is 0.25 M in formic acid (HCO_2H) and 0.50 M sodium formate (HCO_2Na).

3.10 References

Baird, J. K. (1999). "A Generalized Statement of the Law of Mass Action," *J. Chem. Ed. 76*, 1146.

Brice, L. K. (1983). "Le Chatelier's Principle: The Effect of Temperature on the Solubility of Solids in Liquids,"*J. Chem. Ed. 60*, 387.

Chiapple, C., Malaldi, M., Melai, B., Fantini, S., Bardi, U. and Carporali, S. (2010). An Usual Common Ion Effect Promotes Dissolution of Metals Salts in Room-temperture Ionic Liquids: A Strategy to Obtain Ionic Liquid Having Organic-inorganic Mixed Cations, *Green Chem. 12*, 77.

de Levie, R. (2000). "What's in a Name?"*J. Chem. Ed. 77*, 610.

Harris, D. C. (2003). *Quantitative Chemical Analysis*, 6rd ed. W. H. Freeman and Company, New York NJ.

Harris, D. C. (2005). *Exploring Chemical Analysis*, 3rd ed. W. H. Freeman and Company, New York NJ.

Harris, T. M. (1995). "Revitalizing the Gravimetric Determination in Quantitative Analysis Laboratory,"*J. Chem. Ed. 72*, 355.

Harwood, S. W., Herring, F. G., Jeffry, M. D. and Petrucci, R. H. (2012). General Chemistry Principles and Mordern Applications 9th ed., New Jersey, Prentice Hall.

Hill, J. O. and Magee, R. J. (1988). "Advanced Undergraduate Experiments in Thermoanalytical Chemistry," *J. Chem. Ed. 65*, 1024.

Sawyer, A. K. (1983). "Solubility and Ksp of Calcium Sulfate: A General Chemistry Laboratory Experiment,"*J. Chem. Ed. 60*, 416.

Thompson, R. Q. and Mhadiali, M. (1993). Gravimetric Determination of the Molar Mass, *J. Chem. Ed.* J. K. *70*, 170.

Chapter 4

SEPARATION IN ANALYTICAL CHEMISTRY

4.1   Separation

Separation is essential when chemical species in the sample solution interfere with the measurement of the desired constituent. The presence of interfering species results in decrease of the accuracy of the measurement or even hinders a particular analytical method to be used. Separations involve a physical isolation of the desired constituent from the interfering substances. The desired constituent may either remain with the original sample solution or it may be separated from the solution. In order to achieve the latter, chemical reactions are used to create a new phase or to transform one chemical into another. An understanding of the fundamentals and applications of these separation procedures provides a wide range of separation techniques and a subsequent accurate measurement of desired constituent in a new situation. This discussion will focus on some separation procedures, which include:

4.2   Separation by Precipitation

This involves the formation of an insoluble material containing either the desired constituent or the interfering substance. Consequently, filtration can be used to separate the desired constituents from its interference. The desired constituent can remain in solution or in the precipitate. Analyte can be separated by using precipitation method from a mixture or it can be used to remove an interferent from a mixture before analysis. A good thing is that many analytes or interferents can be separated from each other in this way. A few examples in Table 4.1 show the precipitating agents and the ions they precipitate. Indeed, more details of precipitants are thoroughly discussed in section 4.2.1. When separation is a major goal, it is imperative for separation to be *quantitative,* that is, it should be complete. Separation by precipitation requires excess addition of precipitating agent to the solution (under which it avoids colloidal formation and enhances crystallization without occlusion), filtering the precipitate from the reaction mixture, and washing the precipitate free of the precipitating solution. For the case where analyte remains in solution, all the wash solution must be returned to the analyte solution. If the analyte is in precipitate form, it must be retained in the filtering and washing step.

Table 4.1: Some Useful Precipitation Reagents

Reagent	Ion Precipitated
H_2PtCl_6	K^+, Rb^+, Cs^+
$BaCl_2$	SO_4^{2-}
$MgCl_2$	PO_4^{3-}
8-Hydroxyquinoline	Al^{3+}, Fe^{2+}, Mg^{2+}, many more
Dimethylglyoxime	Only Ni^{2+} with controlled pH

Hillebrand, W. F., Lundell, F. G. E., Bright, H. A. and Hoffman, . J. I. *Applied Inorganic Analysis.* John Wiley and Sons, New York, (1953).

The nucleation and crystallization process as will be seen in chapter five, section 5.4.4, suggests the procedure desirable for quantitative separation. Usually, addition of precipitating reagent should be slowly added in a relatively dilute solution, or homogeneous precipitation can be used. It should be noted that the amount of precipitant to be added should be carefully considered. This is because there is a concentration of precipitant at which the solubility of the precipitate is a minimum. On the other hand, large excess of the precipitant can cause resolvation through complexation; however, without relatively excess amount a certain amount of the material will not be precipitated.

It should be clearly noted that the salt that does not react in the precipitating system can be added to assist coagulation of any colloidal particle formed. It is a good practice to warm the solution in order to facilitate the rapid exchange of surface ions and to reduce occlusion. It is advisable for the solution to remain warm after precipitation to allow purification of the precipitate. The solubility of most precipitates increases with increase in solution temperatures, so the solution must be cooled before filtration. The cooling should be done slowly to ensure final growth of the precipitate crystal is gradual.

The precipitate is separated from reaction mixture by filtration. The filtering material can be filter paper held in a filter funnel, or ceramic container with a fritted (somewhat porous) bottom. This container is fitted over vacuum flask and a partial vacuum is applied to the catch flask to help in drawing the filtrate through the filter material. The next step is to transfer all precipitate from the reaction vessel with a clean wash bottle. Particles of precipitate clinging to the wall of the reaction vessel must be dislodged with a tool called *rubber policeman*. Thereafter, both the flask and the rubber policeman must be rinsed, with the solution going into the filter container. One must be cautious not to use too much rinse solution because it can cause error from the solubility of the precipitate in the rinse solution. At some point rinse solution may contain a salt or a wetting agent which can inhibit loss of coagulation or creeping of the precipitate up the sides of the titration flask. Quantitaive precipitate transfer is a skill that requires care and practice.

The extent of completeness of separation by precipitation depends on the relationship between initial concentration and its solubility in the reaction mixture. Suppose the solubility is 1×10^{-6} M for the case of AgCl, a 1×10^{-4} M of Ag^+ or Cl^- would be only 99% separated by this approach. Nevertheless, all analyte concentration above 1×10^{-3} M would be quantitatively separated.

Precipitation reaction can be selectively done by controlling the pH of solution. This is due to hydrolysis of the cation or the anion, as it is the case for complexation reactions. Also the addition of complexation agent that is selective for an interferent can be helpful as well.

The choice of the best precipitant for a certain separation is important and depends on a number of factors. Just to mention a few of these factors: the nature of constituent of interest, the physical characteristics of the resulting precipitate, the solubility, stability and selectivity of the precipitant, as well as its availability. It is true that the entire mentioned variables must be under consideration, though the most important of these factors are the physical characteristics of the precipitate and the selectivity of the precipitant.

It is apparent that often a good separation is achieved by the quantitative formation of an insoluble product. For effective separation, this insoluble product (separable from the precipitating solution by the usual techniques of filtration or centrifugation) must (i) contain at least 99.9% of the constituent of interest, and (ii) not contain a detectable amount of the interfering substance (impurity or impurities). If precipitates are formed out of the interfering substance, one should be able to remove it quantitatively by this procedure and make sure there is no loss of desired constituent through co-oprecipitation.

The above can only be achieved when selective reagent which forms insoluble precipitates in easily separable form is used. However, it should be noted that the selectivity of the reagent is an important aspect which is good for separation. To some extent the term selectivity may be confusing, and general statements concerning it are difficult to make. It is important to remember that any class of precipitating agents may cover a wide range from specific to essentially nonselective reagents which may react with several different ions.

4.2.1 Inorganic Precipitants

Inorganic precipitants include acids, bases and salts. Thus, HCl reacts with Ag^+, Pb^{2+}, Hg^+ ions to form insoluble chlorides. Sulfuric acid precipitates insoluble Ba, Ca, Pb and radium sulfates. Phosphoric acid, H_3PO_4, precipitates or forms phosphates with Al^{3+}, Bi^{3+}, Ca^{2+}, Cr^{3+}, Pb^{2+}, and Th^{2+}, thorium, which are extremely insoluble in water. Hydrogen sulfide, H_2S, forms insoluble sulfides with many metal ions (Sb^{3+}, As^{3+}, Bi^{3+} Cd^{2+}, Cu^{2+}, Ge^{4+}, Pb^{2+}, Hg^{2+}, Pd^{2+} and Sn^{2+}. Inorganic bases, both the alkali hydroxides and ammonia, form insoluble hydrous oxides with many metal ions. For example, ammonia forms complex ions of the type $Ag(NH_3)_2^+$, $Co(NH_3)_6^{3+}$, $Cu(NH_3)_4^{2+}$, and $Zn(NH_3)_4^{2+}$, which are soluble species in ammonical solution, may help in the separation procedure. From the fact that Al (III) and Zn (II) hydroxides are soluble in excess base to give AlO^{2-} and ZnO_2^{2-} ions such characteristic can be useful in separation procedure.

It is apparent that every qualitative analysis student will always remember that AgCl is soluble in aqueous ammonia, and lead chloride is insoluble in the same. With that concept in mind, one can separate silver and lead through first precipipating both metals as chloride and then dissolving the AgCl in aqueous ammonia. The fact that $PbCl_2$ is quite soluble in hot water offers another technique for separating lead from insoluble silver and/or mercury halides.

The solubility of a metal salt containing an anion of weak acid is dependent upon pH conditions of the aqueous medium. For a salt of a monoprotic acid, HA, four equilibrium reaction equations are necessary to describe them (equitation 4.1- 4.4).

$$K_a = \frac{(H^+)\,(A^-)}{(HA)}$$

$$(4.2)$$

$$K_h = \frac{(H^+)\,(OH^-)}{(M^+)}$$

$$(4.4)$$

In acidic solutions, however, the extent of hydrolysis will be very small and can be neglected. Under such conditions, H^+ contributed by the dissociation of water will be negligible, and we can therefore ignore equation 4.3. Let Ca represent the total concentration of the anion A⁻ in solution (free or bound) and α be the fraction of it in the form of the free anion.

$$Ca = (A^-) + (HA), \text{ and} \qquad (4.5)$$

$$\alpha = \frac{A^-}{Ca} = \frac{(A^-)}{(HA) + (A^-)} \qquad (4.6)$$

$$\alpha = \frac{A^-}{Ca} = \frac{(A^-)}{(HA)+(A^-)}$$

Substituting the value of (A⁻) into equation 4.2, expression 4.7 is obtained as:

$$\alpha = \frac{(A^-)}{Ka + (A^-)} \qquad (4.7)$$

Since $(A^-) = \alpha Ca$ from (4.6), equation (4.1) becomes $Ksp = \alpha Ca\,(M^+)$ $\qquad$ (4.8)
Let us apply the above equilibria to calculate the effect of pH on the solubility of a salt.

Example 4.1
Calculate the solubility of strontium fluoride (a) in pure water, and (b) in 0.1 M HF solution, being given Ksp of SrF_2 is 2.7×10^{-9} and Ka of HF is 7.4×10^{-4}. Ignore activity coefficients.

Solution
(a) Let x be the solubility in moles. Litre⁻¹, then $(Sr^{2+}) = x$ and $(F^-) = 2x$
$Ksp = (x)(2x) = 2.7 \times 10^{-9}$
$4x^3 = 2.7 \times 10^{-9}$

$$x = \sqrt[3]{\frac{(7.9 \times 10^{-9})}{4}}$$

$x = 8.8 \times 10^{-4}$ moles/litre
(b) Let us first calculate α from equation 4.7

$$\alpha = \frac{7.4 \times 10^{-4}}{7.4 \times 10^{-4} + 0.1} \approx \frac{7.4 \times 10^{-4}}{0.1} = 7.4 \times 10^{-3}$$

Here, if α is the solubility, the total concentration of the fluoride ions (F⁻ + HF) will be $2x$ and the concentration of the free fluoride ions will be $2\alpha x$. As before, $Ksp = (x)(2x)^2 = 2.7 \times 10^{-9}$ or

$$4x^3 = \frac{2.7 \times 10^{-9}}{\alpha^2} = \frac{2.7 \times 10^{-9}}{(7.4 \times 10^{-3})^2} = 4.93 \times 10^{-5}$$

$4x^3 = 2.7 \times 10^{-9}$

$$x = \sqrt[3]{1.23 \times 10^{-5}}$$

$x = 2.3 \times 10^{-2}$ and as can be seen, the solubility has increased by about a factor of 30.

4.2.2   Organic Precipitants

Organic precipitants that form insoluble compounds with metal ions are important analytical reagents. Many times organic precipitant involves consideration of selectivity, solubility, and cost. Of these, selectivity is of primary concern, and in general, we find the selectivity of an organic precipitant is greatly influenced by the pH of the reacting solution.

All organic precipitants are acidic in nature and must undergo dissociation in solution to yield an anion, which subsequently reacts with a metal ion to form neutral compound. This dissociation involves the loss of one or more protons from organic molecule. Thus, metal ions and the hydrogen ion are competing for ligand. The lower the hydrogen ion concentration, the higher are the chances for the formation of metal anion complex. Therefore, most organic precipitants are more sensitive to pH conditions in the medium than are inorganic precipitants.

Insoluble salts are formed in the reaction of many metal ions with carboxylic acids
R-COOH. Commonly used polyprotic acids in analytical chemistry include citric, oxalic, phthalic and tartaric acid (see Figure 4.1). Insoluble citrates (Ce^{3+}, Cu^{2+}, Hg^{2+}) are known; only oxalic acid is a common precipitant.

Figure 4.1: Some Common Polyprotic Acids

Ligands containing two or more nitrogen, oxygen or sulfur donor atoms or combination of these in position such as five or six-membered chelatering structures can be formed with metal ions, and are the most useful of all organic precipitants. In Figure 4.2 are some examples.

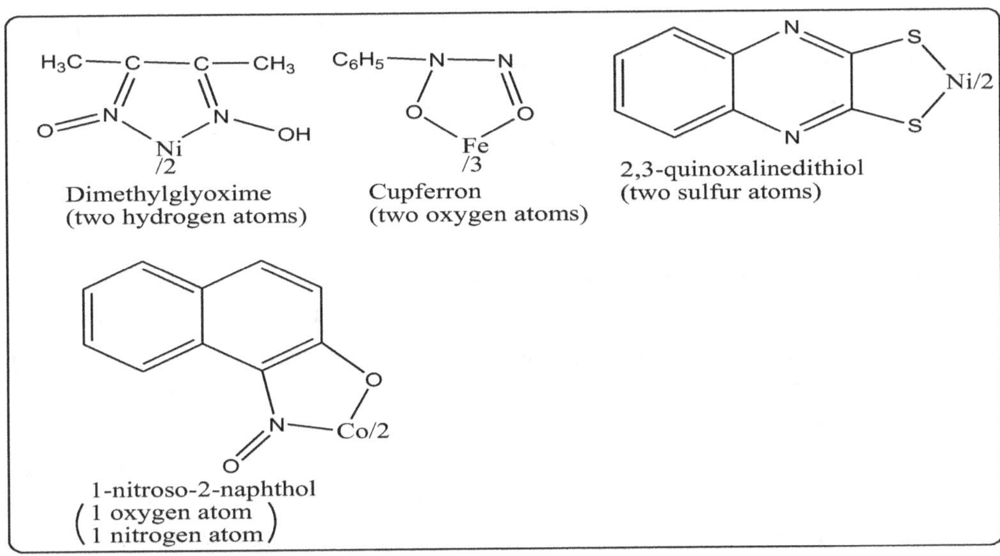

Figure 4.2: Some Ligands Containing Two or More O, N, or S Donor Atoms

4.3   Separation by Electrolysis

Separation of chemical species using electrochemical cells is among the most important methods. Such separation techniques are restricted to the deposition of solid substances on the metallic electrodes and all the reactions taking place at such electrode surfaces in electrolyte solutions should be carefully considered.

Before we proceed, it is imperative that certain definitions and conventions are clearly understood. *Electrolysis* is defined as the decomposition of a chemical compound in solution by passing an electric current. The latter phenomenon occurs in an electrochemical cell consisting of two electrodes immersed in an electrolyte solution. Basically, electrochemical cells are of two types: galvanic (or voltaic) cell in which electrode reactions occur spontaneously, and electrolytic cell in which reactions are driven by an externally applied voltage.

The reaction taking place in electrochemical cell at *cathode* is reduction, and the reaction taking place at *anode* is oxidation. Those two positions are electron transfer reactions, with electron consumption in the reduction reaction and electron liberation in the oxidation reaction. It is apparent that equal numbers of electrons are involved in both reactions in a given cell, which maintain the electroneutrality of the solution.

The electrodes are always surrounded by electrolyte solution, which contains chemical species in ionic form. However, it is not necessary for the number of positive and negative ions to be equal, but there must be an equivalent number of positive and negative charges. Obviously, in an electrolytic cell the externally applied electromotive force (*emf*) sets up a potential difference between the electrodes with a resultant potential gradient in the solution. Because of the existing potential gradient, the positively charged (cationic) species migrate toward the cathode and negatively charged (anionic) species move toward the anode. This movement of ions in the cell results in electric current to flow through the solution.

It is easy at this point to observe ions traveling through a solution under the influence of applied force. Once this process has been in operation for a short time, the solution surrounding the cathode should contain excess of positively charged cations, and similarly the anode solution should contain an excess of anions. Such an unusual situation of non-electroneutrality can be alleviated, as long as the applied force is sufficiently great, by the reactions which occur at the electrode surfaces. Therefore, the positive charge on cations will be neutralized through the transfer of electrons from the cathode to the ions, with simultaneous and equivalent release of electrons from the negatively charged anions to the anode. Any excess of electrons on the anode will be transmitted through the external circuit to the cathode, which will make the process continuous.

The rate of redox (oxidation-reduction) reaction is the factor which determines the magnitude of the current passing through the cell. *Current* is defined as the rate of flow of electrons-quantitatively, the number of coulombs flowing per unit time (second) in an electric circuit. Current is measured in *amperes* (A), which represent a flow of 1 coulomb per second.

In other ways, coulomb is an ampere-second, and is equivalent to the flow of 6.2427×10^{18} electrons (the number of electrons involved in the reduction of the above amount of silver). The *Faraday* is the quantity of electricity equivalent to the transfer of 6.0238×10^{23} electrons, 96,493 coulombs.
Suppose electrons are flowing into a platinum (Pt) wire dipped into a solution: Figure 4.3.

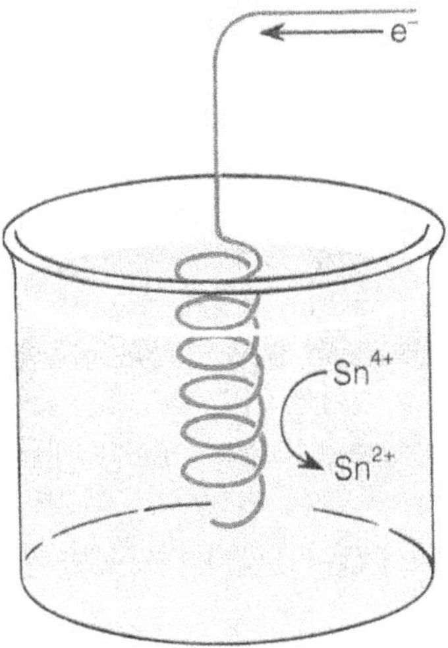

Figure 4.3: Electrons Flowing into a Coil of Pt Wire, Whereby Sn^{4+} is Reduced to Sn^{2+}

The reduction of Sn^{4+} to Sn^{2+} is expressed as can be seen below:

$$Sn^{4+} \quad + \quad 2e^- \longrightarrow \quad Sn^{2+}$$

Basically, Pt as an electrode conducts electrons into and out of chemicals involved in the redox reaction. Platinum being an *inert* electrode has no chance to participate in the reaction except to conduct electrons. The rate of flow of electrons into the electrode measures the rate of reduction of Sn^{4+}.

Example 4.2

Consider Sn^{4+} is reduced into Sn^{2+} at a constant rate of 4.24 mmol/h as in Figure 4.3. How much current flows into the solution?

Solution

We should remember that two electrons are required to reduce Sn^{4+} ion to Sn^{2+}. If Sn^{4+} is reacting at a rate of 4.24 mmol/h, electron flow at a rate of 2 (4.24) = 8.48 mmol/h, which corresponds to:

$$\frac{8.48 \text{ mmol/h}}{3600 \text{ s/h}} = 2.356 \times 10^{-3} \; mmol/s = 2.356 \times 10^{-6} \; mol/s$$

In calculation of the current, the Faraday constant is used to convert moles of electrons per second into colombs per second:

$$\text{current} = \frac{Coulombs}{Second} = \frac{moles}{second} \cdot \frac{coulombs}{mole}$$

$$= \left(2.356 \times 10^{-6} \frac{mole}{s}\right)\left(9649 \times 10^{4} \frac{C}{mol}\right) = 0.227 \; C/s = 0.227 \; A$$

There are three possibilities which chemical species in an electrolytic oxidation or reduction can experience: (i) deposited on electrode (ii) liberated from the solution in the gaseous state, or (iii) converted to different chemical species. The majority of metallic ions are reduced and deposited on the cathode surface. Good examples are cobaltic oxide Co_2O_3 and lead dioxide, PbO_2, which are two solid substances that can be deposited oxidatively on the anode. It should be clearly noted that *cathode* is the electrode at which reduction takes place, electrons are consumed, cations migrate towards, and it has a positive sign. On the contrary, *anode* is the electrode at which oxidation takes place, electrons are produced, anions migrate towards, and it has a negative sign.

$$2\,Co^{2+} + 3\,H_2O \longrightarrow Co_2O_3 + 6\,H^+ + 2\,e^- \tag{4.9}$$

Commonly, hydrogen is a reduced product that is liberated at the cathode. Conversely, halogens and oxygen are liberated as gases at the anode.

A good example of the conversion of one ionic species to another in an electrolytic cell is the reduction of the nitrate ion, in acidic solution, to ammonium ion.

Such solid deposition on the electrode is a reaction of special interest and utility to the analytical chemist. The process can suitably serve two purposes: (1) the separation of a constituent of interest from solution, and (2) the gravimetric determination of the constituent.

Consider a simple electrolysis in which a pair of inert platinum electrodes is placed in an electrolyte solution and externally *emf* is applied across the electrodes. The applied *emf* can be calculated as follows:

where the above change in E_{anode} - $E_{cathode}$ is the back *emf* of the cell, which is often in opposition with applied *emf*; i is the current through the electrolyte cell; and R is the sum total resistance of the circuit. The source of back *emf* is from the fact that the true galvanic cell is present within the electrolyitic cell. For electrolysis of an acidic solution of copper sulfate between platinum electrodes, the electrode reactions are presented as follows:

$$Cu^{2+} + 2\,e^- \longrightarrow Cu \qquad\qquad \text{Cathodic reaction (reduction)} \tag{4.15}$$

$$4\,H_2O \longrightarrow O_2 + 4\,H^+ + 4\,e^- \qquad \text{Anodic reaction (oxidation)} \qquad (4.16)$$

and the overall cell reaction is

Once the electrolysis begins, just a very small amount of oxygen dissolves in the electrolyte solution. As the electrolysis continues, the solution becomes saturated with oxygen and the platinum cathode is plated with copper. Therefore, a galvanic cell is formed which consists of a copper electrode, and an oxygen electrode in an acidic solution of copper sulfate. The latter cell tends to function with the copper electrode undergoing oxidation as follows:

If one looks at this equation carefully one will see that it is directly opposite to the desired reaction (4.17). This also applies to the *emf* of the cell in being directly opposite to the applied *emf*, and must be overcome in order that electrolysis can proceed.

For analytical purposes it is not enough to have an electrolytic cell which just deposits metal. The main idea, and above all, is to effect a *quantitative separation* of the desired constituent and do this in reasonable time. All these conditions are dependent on the voltage applied to the cell. Thus, one must be able to calculate this voltage in terms of the desired experimental conditions.

As we have seen earlier, with most applied *emf* it is necessary to overcome the back *emf* of the cell, and from a qualitative point of view, there should be minimum requirements for the applied *emf*. In order to become clear with this concept, let us examine the potential of each electrode, putting under consideration the analytical concentrations of the chemical species present. Now consider the electrolysis of a solution containing 0.1 M copper sulfate solution in 0.5 M sulfuric acid as an example. For this case large platinum electrodes will be used with vigorous stirring of the solution. Constant and efficient stirring has the advantage of eliminating any concentration gradient set up in the solution as the result of ionic migration under the influence of the applied potential. In case there are concentrations differences at the two electrodes as a result of the electrolysis, a so-called *concentration polarization* will emerge, which is designated as a back *emf* opposed to the applied potential and which requires the application of a bit higher potential than that calculated from the bulk of concentrations of the electrode species. This aspect of polarization is always avoided in quantitative separation by constant stirring of the reaction mixture.

In case of reduction in which copper (II), Cu^{2+} ion proceeds spontaneously, *emf* of the galvanic cell can be calculated from the Nernst equation: see equation 4.21. Only equilibrium concentrations are used in calculation rather than activity to simplify numerical calculations. It should be noted that this approach leads to only approximately correct result values. The half reactions and potentials involved are represented as follows:

$$O_2 \ + \ 4\,H^+ \ + \ 4\,e^- \longrightarrow 2\,H_2O \qquad\qquad E^0 = +1.299 \text{ v} \qquad\qquad (4.19)$$

$$Cu^{2+} \ + \ 2\,e^- \longrightarrow Cu \qquad\qquad E^0 = +0.337 \text{ v} \qquad\qquad (4.20)$$

The copper electrode potential is given by

$$E_{Cu} = +0.337 + \frac{0.0591}{2}\log\,[0.1] = +0.307 \ v \qquad\qquad (4.21)$$

Oxygen electrode potential can also be calculated (see equation 4.22) by assuming a partial pressure of oxygen as 0.2 atmosphere in the solution (the partial pressure of oxygen in a solution exposed to air is approximately 0.2 atmospheres). The back *emf* of the electrolytic (galvanic) cell is therefore 1.219 *v* - 0.307 *v* = + 0.912 *v*.

$$E_{O_2} = +1.229 + \frac{0.0591}{4}\log\,[1]^4\,[0.2] = +1.219 \ v \qquad\qquad (4.22)$$

The sign of the above *emf* being positive implies that spontaneous reaction has occurred in the cell. Indeed, this means an applied *emf* is in excess by 0.912 *v* of which should reverse the reaction and cause copper deposition at the *cathode,* as well as oxygen to be liberated at the *anode.* It has been experimentally noted that quite enough applied voltage is often required to carry out electrolysis at an appreciable rate. The extra voltage to make electrode reaction proceed is called *overpotential* of an electrode. In short, overpotential results during transition of ionic species in metallic atoms being plated on an electrode and the evolution of gaseous chemical species at the electrode surface. Generally speaking, the overpotential is always small for metals, which form reversible electrodes with their ions. Good examples of metals whose overpotential is practically nothing include cadmium, copper, lead, silver and zinc. In comparison, overpotential for metals such as chromium, cobalt, iron, and nickel differ slightly, and are always with the same order of magnitude as the overpotential of hydrogen evolution. The overpotential for gases is likewise large. The overpotential of hydrogen evolving at a bright platinum cathode is 0.162 v at a current density of 0.01 amp. cm^{-2}. On the other hand, in the same conditions, the overpotential on a mercury cathode is 1.2 v. Also overpotential of oxygen on a smooth platinum anode is approximatelly 0.40 v. It is important to realize that the actual value of the overpotential in all cases is a function of current density (current density per unit area of the electrode), the nature of the electrode, the condition of the electrode surface, and temperature.

The foregoing discussion has revealed the necessity of increasing the applied voltage beyond the already calculated value of +0.912 volts in order to get the required electrodeposition of copper. Because the overpotential for the deposition of copper on a smooth platinum electrode is negligibly small and that for oxygen evolution from a similar electrode is approximatelly 0.40 v. Thus, an applied voltage of 1.31v is necessary for the electrolysis of copper sulfate solution. So far, we have seen important contributions of applied potential on the electrolytic cell. It is time to look at the current characteristic of the cell, because the time required for complete electrolysis is a function of current. Recalling *Faradays Law*, which stated that one faraday of electricity, 96,493 coulombs (amps. sec), is required to oxidize or reduce one equivalent of any substance. The latter fact implies that one ampere would deposit 31.77 grams of copper in 96.493 seconds. To think of this in an analytical perspective, let us imagine that we had attempted the electrodiposition of copper from 100 mL of a solution containing 0.05 M copper sulfate in 0.5 M sulfuric acid. The resistance is almost 5 ohms. Thus, the current under an applied voltage of 2.00 v as it can be seen in equation 4.23.

$$i = \frac{E_{applied} - (E_{anodic} - E_{cathodic})}{R} \tag{4.23}$$

Suppose anodic potential (with overvoltage) is 1.62 v, the cathodic potential is 0.307 v ≈ 0.31 v. Therefore,

$$i = \frac{2.00 - (1.62 - 0.31)}{2} = 0.14 \ apm. \tag{4.24}$$

The current above, 0.14 amp., is just approximate and valid at the beginning of electrolysis. The value of current decreases as the potential of the cathode increases with decreasing copper concentration. It is obvious, however, that 10 millifaradays are required to deposit 5 millimoles of copper. Thus, assume we have 100% current efficiency, 965 coulombs of electricity are necessary for the electrolysis to happen. Suppose the initial current of 0.14 amp flowed throughout the electrolysis (obviously the current decreased exponentially with time), thus, electrolysis is so slow and time consuming.

It should be clear that electrolytic separations are often carried out either at constant current or under controlled potential. Under constant current electrolysis, the applied voltage is reasonable enough to cause the reduction of the required constituent as well as the reduction of some other ions which may be present in much greater concentration. Consider an example of deposition of copper from acidic solution; an applied potential of 2.5 v will results in the simultaneous reduction of copper and hydrogen. A constant current due to hydrogen reduction will be kept constant throughout electrolysis process though the currentcaused by the relatively low concentration of copper, which approach approaches zero as the copper concentration approach zero.

In general, the evolution of hydrogen gas from the cathode causes copper deposit to be rough and poorly adherent; therefore, it is essential to control the cathodic potential by the reduction of some species other than hydrogen ion. For example, nitrate ion can be reduced to a lower potential than hydrogen ion, and most of the time serves as an ideal cathodic depolarizer, which results in decrease of hydrogen evolution, and that decrease happens without the formation of gaseous product as can be seen in equation 4.25.

This means quantitatively that electrodeposition of copper from a nitric-sulfuric acid mixture is, thus, very easy to obtain.

Electrolysis has been carried out under constant current (apparatus setup is shown in Figure 4.4) to separate copper from aluminum, cadmium, cobalt, manganese, and nickel. While nickel can be deposited quantitatively from ammonical solution, lead can be deposited as the dioxide by anodic oxidation under constant current conditions.

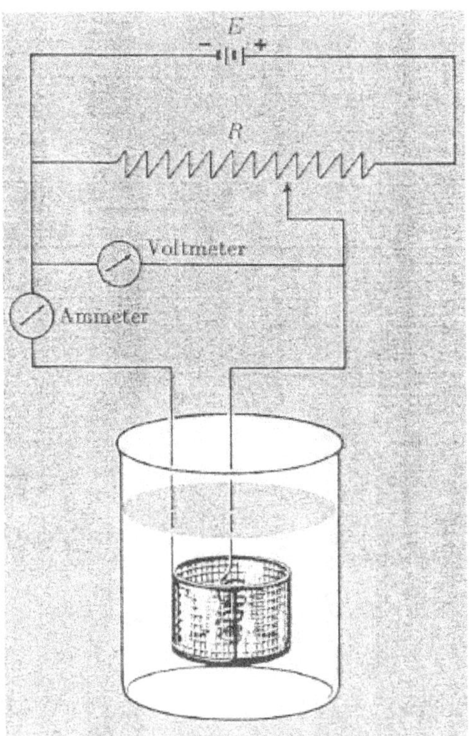

Figure 4.4: Apparatus for Constant Current Electrolysis with Platinum Cathode

A constant current of several amperes applied to a mercury cathode with an applied voltage ranging to 6-12 V, has been commonly used in analytical separation. Cobalt, copper, iron, manganese, and nickel can be reduced and directly deposited on a mercury cathode from a dilute sulfuric acid (H_2SO_4) solution. Surprisingly, under the same conditions, aluminum, molybdenum, titanium, tungsten, and vanadium are not separated from the solution.

It is clear that the selectivity of a reduction is governed by the potential of the cathode. Thus, the cathodic potential controls the reduction reactions occurring in the electrolytic cell. Basically, the potential of the cathode is controlled in constant current electrolysis by decreasing the cathodic depolarizer or hydrogen ion. Indeed, the cathodic potential is such that many selective separations are not allowed.

Under controlled potential electrolysis (apparatus setup Figure 4.5), the use of a third electrode allows the measurement of the cathode (or anode) potential during the electrolysis. When electrolysis is in progress there is a chance of holding cathodic potential constant by decreasing the applied potential either electronically or manually. The use of that technique has enabled selective separations of antimony from tin, bismuth from copper, cadmium from zinc, and copper from silver. Indeed, electrolysis is commonly used in the manufacturing of many important chemicals (sodium, chlorine, sodium hydroxide, aluminum) and in numerous processes for purification and electroplating of metals. The latter process is the coating of one metalon the surface of another using electrolysis.

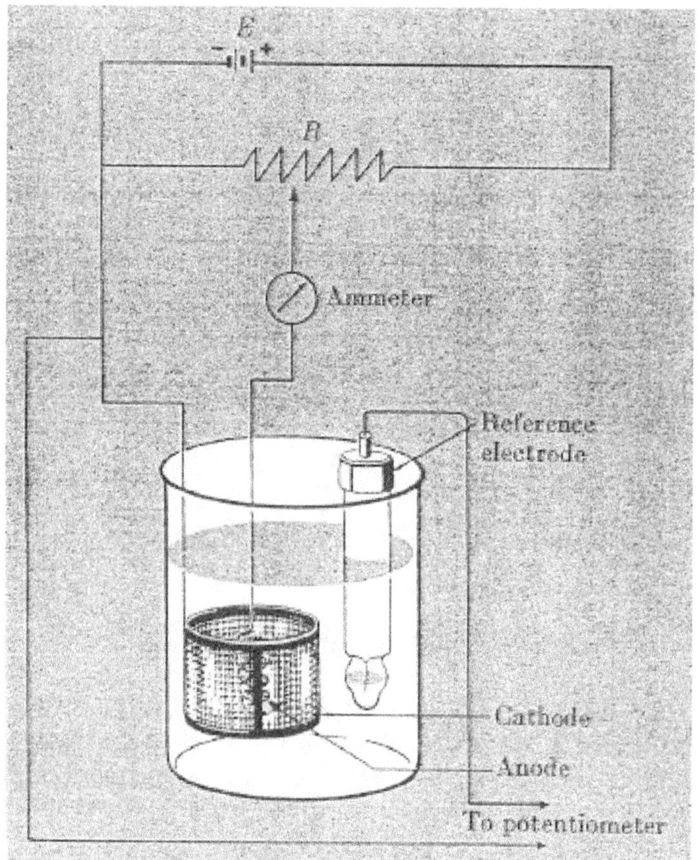

Figure 4.5: Apparatus for Controlled Electrolysis

4.4 Separation by Ion Exchange

Ion exchange is a reversible exchange of ions between a solid phase and a liquid phase which takes place in an equilibrium controlled process. Cation exchangers have overall negative charged groups on stationary phase that attract *cations* in a given solute. Conversely, anion exchangers have overall positive charged groups on stationary phase that attract *anions* in a given solute. For the purpose of understanding and appreciation of the ion exchange process, we should look at the nature of the resin particle used in the column during separation process. Resin(s) are usually in spherical beads, is a three dimensional, randomly, hydrocarbon polymer containing a large number of ion exchange sites. The length of the polymer chains and the cross-linking between chains are great enough to ensure insolubility in water, but low enough to allow permeation by the electrolyte solution containing the ions to be separated. Two characteristics of ion-exchange resins, particle size and cross-linking, have important consequences.

(i) *Particle Size*
Particle size influences equilibrium time and the efficiency of a resin. If the resin particle is in a salt solution, an exchange of hydrogen ions for other cations takes place. Small size particle not only decreases equilibrium time, but also increases the efficiency of a given volume of resin since there is better contact between the solution and the exchange sites. In addition, decreasing particle size increases pressure drop in an ion exchange column, and thus decreases the flow rate of solution through the column.

(ii) *Degree of Cross-linkage*
Degree of cross-linkage in a resin has far more effect than does particle size. Cross-linkage refers to the percentage of divinyl benzene used in preparation for a copolymer.
For example, Figure 4.6

Figure 4.6: Copolymer Fragment

The degree of cross-linkage influences and affects the following:
(i) The swelling characteristics of the resin particle (ii) the capacity of resin (iii) equilibration rate, i.e., the time necessary for establishment of equilibrium conditions, and (iv) selectivity.

It should be remembered that highly cross-linked polymer is not free from swelling in a solution. As a result, diffusion of ions into the intricate polymer network is slow. Equilibration time, therefore, is increased with increased cross-linkage; selectivity is also increased because certain large ions cannot diffuse through a very fine polymer network. Therefore, discrimination between the resin pore sizes shrinks as cross-linking increases. The following are some important technical terms in separation. *The capacity of an ion exchange resin*: This is the number of ionic exchange sites per unit weight or volume of resin.
The wet volume total capacity: This is the number of exchange sites per unit volume of the swollen resin.

The selectivity of an ion exchange: This is determined by the affinity of the ion exchange resin for a particular ion, and experimental conditions under which ion exchange occurs.

Ion exchange affinity: This is the degree of attraction between an exchange site on the resin and an ion in solution, and is independent upon the size and the charge of the ion.

The rules governing ion exchange selectivity have been formulated by Kunin as follows:
(a) ion exchange affinity increases with increasing ionic charge, for example, the following is the order of increasing affinity: $Na^+ < Mg^{2+} < Al^{3+} < Th^{4+}$. For ions with *like* charge, ion exchange affinity increases with increasing atomic number. In the case of alkali metals we have: $Li^+ < Na^+ < K^+ < Rb^+ < Cs^+$; (b) differences in ion exchange affinities decrease with increasing ionic concentration; and (c) the affinity of an exchanger for hydrogen (H^+) and OH^- ions is dependent upon the nature of the exchange group in the resin. For strongly acidic cationic resin, the affinity for H^+ ion is very low: a similar relationship exists for OH^- ion with anionic resins.

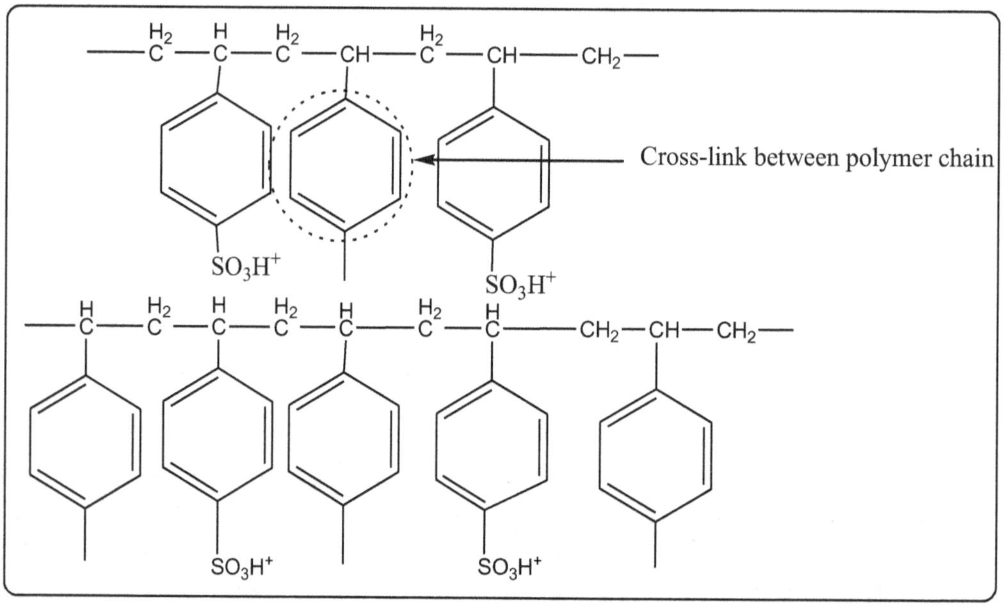

Figure 4.7: Structure of Polystyrene Cation Exchange Resin

There are two examples of resin used in ion exchange chromatography, which include polystyrene, which consists of amorphous (noncrystalline) particles. Polystyrene is basically made into a cation exchanger while sulfonate ($-SO_3^-$) or carboxylate ($-CO_2^-$) groups are attached to the benzene rings. Figure 4.7 following is an example of polystyrene *cation-exchange* resin. Cross-links are covalent bridges between the polymer chain.

Figure 4.8: Structural Example of Polystyrene *Anion - exchange* Resin

Polystyrene is an anion exchanger if ammonium groups ($-NR_3^+$) are attached, as can be seen in Figure 4.8.

Another example of anionic exchanger includes aluminum oxide (Figure 4.9), which strongly adsorb anions including HSO_4^-, F^-, and $H_2PO_4^-$, while $Cr_2O_7^-$ is moderately strongly adsorbed. These characteristics are usually important in the determination of the amount of sulfur in steel. The bisulfate ion has good adsorption in a hydrochloric-perchloric acid solution and therefore enables separation from Fe^{3+}, which is always a source of trouble in the determination of sulfur as $BaSO_4$:

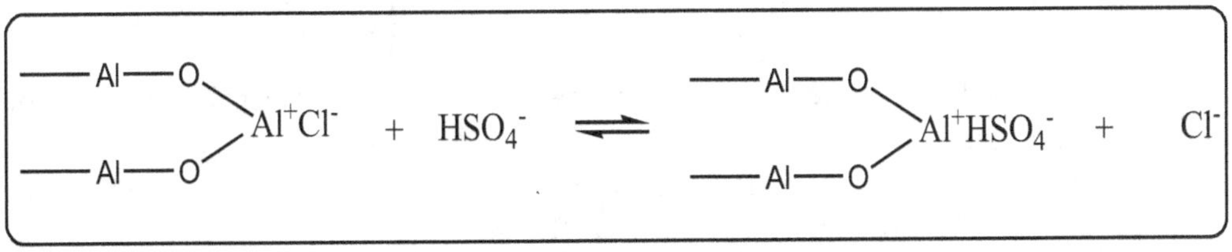

$$\text{Al-O}\diagdown\text{Al-O}\diagup\text{Al}^+\text{Cl}^- + HSO_4^- \rightleftharpoons \text{Al-O}\diagdown\text{Al-O}\diagup\text{Al}^+\text{HSO}_4^- + Cl^-$$

Figure 4.9: Aluminum Oxide as Anion Exchanger Resin

The sulfate in the resin is eluted by addition of ammonia solution, where HSO_4^- is replaced by OH^-. Indeed, upon washing the column packed with aluminum oxide with an alkali solution, this column can function as a cation exchanger. This tendency has allowed lithium to be easily separated from other alkali metals by adsorption on aluminum oxide at pH 12.6.

4.4.1   Factors Influencing Extent of Ion Exchange

The adsorptions of ions by resins are due to a number of factors, among which the following are considered to be important.

Nature of the ions: If all factors remain the same, the higher the charge of ion, the more strongly it is adsorbed. The higher oxidation state rare earth metal ions have higher adsorption than calcium by a cation-exchange resin; on the other hand, calcium is more strongly adsorbed than alkali metal ions.

Size of hydrated ions: Adsorption increases with decreases in solvated ionic volume. This is the trend for the increasing adsorption for the alkali metals: $Li < Na < K < Rb < Cs$. For example, sodium can be separated from potassium much more easily by ion exchange chromatography because potassium is relatively more strongly adsorbed than sodium on the same cation resins. Generally speaking, adsorption increases as the hydration of an ion decreases and its polarizability increases. Relatively large organic ions are strongly adsorbed, nevertheless, and if they are excessively too large may not be able to enter the resin structure. In addition, it should be clearly noted that strong-base anion exchangers show strong affinity for complex anions.

Composition of the solution: The adsorption between two different ions may depend to some extent on their concentrations in the solution. The concentration of substances combined with ions to form partly dissociated uncharged molecules or complex ions in solution will have a huge effect on adsorbability. A good example is cobalt, which does not get adsorbed at all from 1.00 M hydrochloric acid by strong-base anion exchange resins, whereas it is strongly adsorbed from 8-9 M hydrochloric acid in which anionic cobalt chloro complexes are formed.

4.4.2   Rate of Exchange

This can be perceived from the following reaction equation:

$$RX + Y^+ \rightleftharpoons RY + X^+$$

There are three basic factors which can be used in the determination of the rate of exchange:

1. Diffusion of Y^+ to the surface of the resin particle and simultaneous diffusion of X^+ from the surface into the solution.
2. Diffusion of Y^+ into the resin to the exchange site and diffusion of X^+ out of the resin.
3. Replacement of X^+ by Y^+ at the exchange site.

It appears that the third factor is taking place very rapidly and it cannot be considered as limiting in some chelating exchangers. Thus, the exchange rate is diffusion-controlled, and, depending on either factor number 1 or 2, may be the rate determining. The limiting factor in step 1 is the diffusion of the ions through the "Nernst film," a stationary film existing at the surface of the particles even when the bulk of the solution is well mixed, but on keeping raising in agitation the bulky staff diminishes? . The rate of film diffusion increases with the raise in temperature. The degree of cross-linking in step 2 above has an important role to play in the resin. Conversely, in low cross-linking, swelling is relatively higher and porosity is increased so that the diffusion rate is increased. For the case of high cross-linking, where channels are more twisted and make a smaller fraction of the volume, the diffusion coefficient in the resin may be significantly lower than at low cross-linking.

4.4.3   How Ion Exchange Works

In order to explain the concept more clearly let us use water as an example. Ions are atoms, small particles that are the building block for molecules. Ions have a weak electrical charge (positive (+) for cation or negative (-) for anion). For example, positively charged Na^+ ions are commonly used to coat cation exchange resins. Negatively charged chloride Cl^- or OH^- are commonly used to coat anion exchange resins. Mixed bed resins combine both positive and negative ions.

Ion exchange units actually can exchange ions from resins with those in water. When water to be treated passes through the ion exchange unit, ions in the water are attracted by either a positive or negative charge to the resin bed. Since the ions from the water are usually held more tightly by the resins than they were held in the water, they are, in effect, removed from the water in the exchange process. The latter technique has been very useful in analytical chemistry to measure ions in cloud droplets. This can be seen in Table 4-2, which shows the mean composition of non-precipitating clouds for six summer months in 1996 at the summit of Mt. Brocken in the Harz Mountains of Germany.

4.4.4   How Cation Exchange Units (Softeners) Work

As we have seen previously, cation exchange resins are usually coated with positively charged Na^+ ions. When water containing dissolved cations get into contact with resin, the cations are "exchanged" for or trade places with the loosely held sodium ions on the resin. In this way, the Ca^{2+} and Mg^{2+} ions which are responsible for hardness are removed from the water and placed on the exchange resin, and the Na^+ from the resin is added to the water. This process makes the water 'soft'.

Table 4.2: Ionic Composition of Clouds

Ion	Concentration (μM)	Ion	Concentration (μM))
Cl⁻	101	H⁺	131
NO_3^-	360	Na⁺	100
SO_4^{2-}	156	NH_4^+	472
		K⁺	1.3
		Ca^{2+}	26
		Mg^{2+}	12

Data from Acker, K. et al. (1998). *J. Anal. Chem., 360, 59.*

The existence of calcium (Ca^{2+}) and magnesium (Mg^{2+}) in water are considered to be a source of hardness, and are reported in milligrams per litre (mg. L^{-1}), or parts per million (ppm). It is approximated that 4.5 mg of sodium are enough to be added per litre of water to reduce hardness. There is a point at which sodium ions remain on the resin; this implies no more calcium or magnesium ions can be removed from the incoming water at this point. At this stage the resin is considered *exhausted* or spent and can no longer accomplish further water treatment until it is *recharged* or *regenerated*. The latter can be done by washing the resin with sodium carbonate solution. It is important to note that not all water from different parts of the world has enough minerals to require softening.

Generally speaking, *water hardness* means the total concentration of alkaline earth metal ions in that water. Often, the concentration of both Mg^{2+} and Ca^{2+} are greater than any other Group II ions; thus, hardness can be equated to $[Mg^{2+}] + [Ca^{2+}]$. Commonly, hardness is expressed as the equivalent number of milligrams of $CaCO_3$ per litre. Therefore, $[Mg^{2+}] + [Ca^{2+}] = 1$ mM; alternatively, one can say that hardness is 100 mg per litre, due to the fact that 100 mg $CaCO_3$ = 1 mmol $CaCO_3$. Conversely, water, it's the hardness of which is less than 60 mg $CaCO_3$ per litre, is often considered as soft. The reaction of soap and hard water always results in the formation of insoluble curds as follows:

A reasonable amount of soap must be available to consumed Mg^{2+} and Ca^{2+} before the soap is useful for cleaning. On the other hand, hardness is beneficial in irrigation water because the presence of alkali earth metal ions leads to the formation of *flocculate* (cause aggregate) colloidal particles in soil, which increase the permeability of the soil to water.

4.4.5   What Does Cation Exchange Remove?

Water softeners exchange calcium (Ca^{2+}) and magnesium (Mg^{2+}) with sodium (Na⁺) ions. Such exchange happens as the hard water passes through a resin bed which attracts and holds tightly Ca^{2+} and Mg^{2+} in exchange with Na⁺. Basically, both Ca^{2+} and Mg^{2+} concentration if exceeding certain levels can scale pipes, water heaters, boilers, and appliances, lowering water flow rate and efficiency. Cation exchange resins also remove barium (Ba^{2+}), cadmium (Cd^{2+}), copper (Cu^{2+}), manganese (Mn^{2+}), radium (Ra^{2+}), zinc (Zn^{2+}), and other metallic, positively charged ions.

4.4.6 How Anion Exchange Works

Anion exchange units have a resin that exchanges Cl^- or OH^- for the anions that they remove. What does anion exchange remove? Anion exchange units can remove nitrate, (NO_3^-), sulfate (SO_4^{2-}), and other negatively charged anions. Typical examples of ions that can bind to ion exchangers are: H^+ and OH^- single charged monatomic ions like Na^+, K^+ and Cl^-, double charged monatomic ions like Ca^{2+} and Mg^{2+}, polyatomic inorganic ions like SO_4^{2-}, and PO_4^{3-} organic acids, usually molecules containing the amino functional group $-NR_2H^+$ organic acids, and usually molecules containing -COO- (carboxylic acid) functional groups.

4.4.7 Advantages and Disadvantages of Water Softener

Usually the cost of water softener is well balanced with the savings of soft water. It has been reported from the literature that water softening reduces the quantity of cleaning products by almost fifty per cent. In addition, the lifespan of plumbing systems and water-using appliances can be relatively extended. It is time-saving, particularly in cleaning and removing scale and gives better results in laundry and dishwashers. Moreover, water softener can cause heath effects by removing beneficial calcium and magnesium and substituting sodium. Most of the time sodium added to make water soft is a relatively small fraction of the sodium intake obtained from other sources and probably may not be the case for healthy people. Nevertheless, people restricted to taking a certain amount of salts in their diets are urged to consult their physician before deciding to use soft water from ion exchange units for both cooking and drinking.

4.4.8 Major Applications of Ion Exchange

Ion exchange is commonly used in the food and beverage, hydrometallurgical, metal finishing, chemical, petrochemical, pharmaceutical, sugar, ground and potable water, softening and industrial water industries.

4.4.9 Applications of Ion Exchange in a Few Cases

(i) *Deionized water*
What is deionization?
This is simply the removal of ions in water. Ions refer to electrically charged atoms or molecules found in water with either a net negative or positive charge. Typically, ion exchange resins are used to exchange non-desirable cations and anions with hydrogen and hydroxyl, respectively, forming pure

water (H_2O), which is not an ion. Table 4.3 shows ions commonly found in most of the world's municipal waters.

Table 4.3: Common Ions Found in Municipal Water

Cations	Anions
Removed by Cation Resins	*Removed by Anios Resins*
Calcium (Ca^{2+})	Chloride (Cl^-)
Magnesium (Mg^{2+})	Sulfate (SO_4^{2-})
Iron (Fe^{3+})	Nitrate (NO_3^-)
Manganese (Mn^{2+})	Carbonate (CO_3^{2-})
Sodium (Na^+)	Silica (SiO_2^-)
Hydrogen (H^+)	Hydroxyl (OH^-)

Example 4.3

How is deionized water prepared?

Solution:

This is prepared by allowing water to pass through an anion - exchange resin loaded with OH^- and a cation - exchange resin loaded with H^+. Suppose, for example, $Cu(NO_3)_2$ is present in the water. The cation - exchange resin binds Cu^{2+} and replaces it with 2 H^+. The anion - exchange resin binds NO_3^-, and replaces it with OH^-. The H^+ and OH^- combine, so the eluate is pure water, as can be seen in the short summary description below.

The exchange occurs in the production of deionized water because of the fact that ions are attracted to resin beads with different strengths. For example, Ca^{2+} ion is more strongly attracted to cation resin beads than Na^+ ion. Both H^+ on the cation resin beads and OH^- on anion resin beads do not have a strong attraction to the beads. Essentially, this is what allows ion exchange to take place.

Example 4.4

How can ions in water be measured?

Solution:

Electric current passes through water using ions as stepping-stone to pass current from one conductive ion to another. Therefore, by measuring the electrical conductance of water can always tell us the ionic contents of water. Obviously, fewer ions in the water will make the passage of electricity more difficulty. Thus, water with a lower conductivity value is more deionized than water with a high conductivity value.

(ii) *Preconcentration*

Measuring extremely low levels of analyte, i.e. trace analysis. Trace analysis is important for environmental problems in which low concentrations of substances, such as mercury in fish, can be concentrated over many years in people who eat a large quantity of fish. For trace analysis concentrations are always low and cannot be measured without preconcentration, a process in which analyte is brought to a higher concentration prior to analysis. Metals in natural waters can be preconcentrated with a cation - exchange column, a cross sectional of column can be seen as follows:

Once a large volume of water is passed through a small volume of resin, then the cations are concentrated into a small column. Thereafter, the cation can be displaced into a small volume of solution by eluting the column with concentrated acid. This is now concentrated enough to be easily identified by using an analytical instrument, for example, Atomic Absorption Spectroscopy (AAS) spectrometer.

4.5   Separation by Solvent Extraction

Extraction as a technique means the action of taking something out from its environment or medium. Typically, solvent extraction is the most widely used in analytical separation, though there are various extractions in which matter other than liquid could be used.

In *liquid-liquid* extraction there are two solvent phases, which are immiscible or slightly miscible in each other and a solute distributed between the two phases. The actual chemical compound distribution between two phases is equilibrium controlled and independent of the total solute concentration. Suppose we have a certain compound B in volume of solvent 1 which was shaken with a known volume of solvent 2 in a separatory funnel, obviously some of compound B will migrate into the new solvent. Consequently, the equilibrium will be established, and at that time the rate of migration of B from solvent 1 to solvent 2 will be equal to the rate of reverse migration.

If we assume homogenous chemical equilibrium expression 4.26, it is correct to write an equilibrium constant expression as follows in equation 4.27:

$$K_D = \frac{(B_2)}{(B_1)}$$

(4.27)

where K_D is the *distribution coefficient*. Indeed, this is a true equilibrium constant and represents the ratio of activities of the compounds in two phases. Because we are dealing with molecular species rather than with individual ions, activities can be replaced with concentrations with minimum possible error.

The distribution coefficient is commonly used when it is clearly known that the same species of a compound are present in two phases. For the case when in one phase the compound undergoes chemical reaction to produce more than one chemical species, for example, dissociation of acid into ionic species, there will be distribution effects. Expression 4.28 accommodates all varieties of solute of great value than just expression 4.27 this sentence is not clear. The *distribution ratio*, D, takes care of all species of a solute and is represented as in expression 4.28.

$$D = \frac{Total\ concentration\ of\ solute\ B\ in\ phase\ 2}{Total\ concentration\ of\ solute\ B\ in\ phase\ 1}$$

(4.28)

In addition, if a solute is extracted from aqueous phase into an organic solvent phase, the distribution ratio is represented as in expression 4.29.

$$D = \frac{Total\ concentration\ of\ B\ in\ organic\ phase}{Total\ concentration\ of\ B\ aqueous\ phase}$$

(4.29)

Consider a solute which has both acidic and basic properties: thereby that nature of the species in the aqueous phase will be a function of pH according to the following equilibria in expressions 4.30 and 4.31, respectively.

$$K_1 = \frac{(H^+)\,(HX)}{(H_2X^+)}$$

(4.30)

$$K_2 = \frac{(H^+)\,(X^-)}{(HX)}$$

(4.31)

Experience has shown that only the neutral molecule, HX, will be extracted into an organic extractant. Obviously, individual ions often cannot be extracted because the electrical neutrality of the solutions must be maintained. As has been mentioned, only neutral molecules are extractable into an organic phase, since the solutions must always remain electrically neutral. This means during extraction involving ionic species, the extraction must be preceded by the formation of ion aggregates or uncharged compounds. For the case of metallic cations this can be extracted through the formation of coordination or an ion-associated complex. A good example is copper oxine complex: $Cu(C_9H_6OH)_2$ is a chelate which is extractable from water into chloroform layer as a neutral specie. Through association the charged ionic species are neutralized by certain cations.

Another step in the ionic aggregate extraction process is the distribution of species into two immiscible phases. For such cases the distribution depends on the relative solubilities of extractable compound in the two media. Often, covalent complexes such as the cupferrates, $C_6H_5N(NO)O-NH_4^+$ and sodium diethydithiocarbonamate $(C_2H_5)_2NC(S)S-Na^+$ are more soluble in organic solvents than in water. Likewise, ion-associated complexes, for example MRY, where R is a coordinate ligand and Y is an anion, are often more soluble in a nonaqueous than in an aqueous medium, particularly when R is large organic molecule.

There are three important factors which influence the effective separation of mixture with diverse ions which include the pH of aqueous phase: the reagent used in forming the uncharged extractable species and the type of organic solvent used. Cautious choice of these three variables allows separation in numerous systems.
$$K_2 = \frac{(H^+)(X^-)}{(HX)}$$

$$D = \frac{Total\ \ concentration\ \ \ \ of\ solute\ \ B\ inphase\ \ \ 2}{Total\ \ concentration\ \ \ \ of\ solute\ \ B\ inphase\ \ \ 1}$$

$$D = \frac{Total\ \ concentration\ \ \ \ of\ solute\ \ B\ inphase\ \ \ 2}{Total\ \ concentration\ \ \ \ of\ solute\ \ B\ inphase\ \ \ 1}$$

It is obvious that there are three applicable extractions in the analytical laboratory. These are batch extraction, continuous extraction, and discontinuous countercurrent extraction. Of these three, only batches extraction will be discussed in this chapter.

For batch extraction, a given volume of an aqueous solution containing appropriate solute is shaken with a certain volume of an extracting organic solvent in a separatory funnel until the equilibrium is established, i.e., this process proceeds until there is no more transfer of solute from the aqueous to the organic phase. When the mixture is allowed to stand, the two layers separate and the solute is distributed between the two solvents under equilibrium conditions. Often, solute extraction from the aqueous phase is dependent on the distribution ratio of the two solvents.

Consider the v mL of solution (phase 1) containing w grams of solute A in equilibrium with s mL of solvent (phase 2). Once the equilibration has been attained and separation of the two phases, w_1 grams of solute remain in phase 1. The concentration of A in phase 1 is w_1/v grams per mL, and that in phase 2 is $(w - w_1) / s$ grams per mL as shown in expression 4.32.

$$D = \frac{(w - w_1)/s}{w_1/v}$$

(4.32)

and

$$w_1 = w\left(\frac{v}{Ds + v}\right)$$

(4.33)

The actual amount of solute, w_2 grams, remain in phase 1 after a second extraction with another portion of s mL of solvent is shown in expression 4.34.

$$w_2 = w_1\left(\frac{v}{D_s + v}\right) = w\left(\frac{v}{D_s + v}\right)^2$$

(4.34)

$$w_n = w\left(\frac{v}{D_s + v}\right)^n$$

(4.35)

For the case where n successive extractions with s mL portion of solvent have been carried out, w_n grams of solute will remain in phase 1, as indicated in expression 4.35. It should be noted that the larger the value of partition coefficient and the greater volume of organic solvent used, the larger is the fraction B that gets extracted. It is advised not to use a large volume of extracting solvent, due to the fact that the species extracted may become too dilute.

Example 4.5

In the extraction of solute whose distribution ratio between the two solvents is 50, 1 gram of material in 100 mL of water was extracted with an organic solvent. Calculate the amount of extracted (a) with three 10 mL portions of the extracting solvent, and (b) with one 30 mL portion of this solvent.

Solution:

(a) After one extraction the following is obtained,

$$W_1 = 1 \left(\frac{100}{500 + 100} \right) = 0.16667 \; gram$$

With two extractions the following is obtained,

$$W_2 = 1 \left(\frac{100}{600} \right)^2 = 0.0273 \; gram$$

And after three extractions the following is obtained,

$$W_3 = 1 \left(\frac{100}{600} \right)^3 = 0.0046 \; gram$$

(b) For just a single extraction using 30 mL of extractant (equivalent to the three separate extractions)

$$W_1 = 1 \left(\frac{100}{1500 + 100} \right) = 0.0625 \; gram$$

The above results mean that quantitative extraction is best carried out by multiple extractions with small portions of extractants as opposed to a single extractant. Because of that, batch extraction is a frequently used technique in preparation for the analysis of organic constituents in biological or environmental fluids. To some extent, it is selective for nonpolar compounds, but not that much selective for various nonpolar constituents that may be present.

Consider solute which is dissolving in aqueous phase consisting of a two-component mixture; under such a circumstance a partial separation will be achieved in a batch extraction if the distribution ratios of the two components differ greatly. Suppose the separation factor, C, measures the effectiveness of separation of the two components and is expressed as:

$$C = \frac{D_1}{D_2}$$

where D_1 and D_2 are the distribution ratios for components 1 and 2, respectively. For the purpose of achieving quantitave separation, i.e., 99-100% of component 1 in the presence of 0-1% component 2 in either phase, the minimum value of C must be as follows:

$C = 10^2/10^{-2} = 10,000$.

4.6　Separation by Volatalization

Most volatile substances can be separated from nonvolatile materials by vaporization. Compounds with a high vapor pressure and with a small tendency toward intermolecular attraction in solution can be separated relatively easily. The nonmetallic elements are more volatile than the metallic elements and form many volatile compounds. The rare gases, N_2, O_2, halogens, CO_2, H_2O, SiF_4, SO_2, NH_4Cl, etc. are some examples of volatile elements or compounds, which may be separated by vaporization. Fractional distillation has been used as a technique to separate most of organic compounds with high volatility. Through using advanced techniques of separation, even compounds differing with only 2 ᵒC boiling points can be separated under the proper operating conditions. Volatilization techniques are dependent upon the desired fate of the volatile substance.

For removing interfering substances and for the desired to remain in the sample medium, the volatile substances may be allowed to disperse into the atmosphere or a fume hood, or they may be condensed and discarded. The desired constituent is an integral part of the gaseous product; it must be contained in a closed system and recovered by appropriate means.

Nonmetals forming volatile species which may be separated by vaporization include; boron (B), bromine (Br), carbon (C), chloride (Cl), fluorine (F), iodine (I), nitrogen (N), oxygen (O), selenium (Se), sulfur (S) and tellurium (Te). The following are examples of some volatile species e.g., CH_3BO_2, CO_2, SiF_4, NH_3, N_2, CO, SO_2, H_2S, $SeCl_2$, $TeCl_2$, Cl_2, Br_2 and I_2. It should be clearly noted that vaporization is one of the separation techniques which are rarely used in analytical separation of metals; however, the technique is feasible where appreciable differences in boiling points exist.

4.7 Problems

4.7.1. (a) What is deionized water? What do you think about common impurities, which are not removed by deionization as a process?

4.7.2. How does preconcentration of cations work with an ion exchanger function? Explain. In your opinion, what is the reason why the concentrated acid eluant must be very pure?

4.7.3. (a) As an expert of separation using volatilization, you are given compounds with high vapour pressure, but they have a high tendency toward intermolecular attraction in solution. What are the best options considered relevant in such separation? (b) What is the prerequisite for metals to be separated by voltalization technique?

4.7.4. Provide an explanation of how you are going to remove volatile interfering substances in the desired sample when you are carrying out separation using voltalization technique.

4.7.5. (a) What is liquid-liquid extraction? Explain. (b) Suppose you have a mixture of acidic, basic, and neutral species which you are trying to separate by liquid-liquid extraction. Which of those species do you expect will be found in the organic layer?

4.7.6. (a) Why must species be electrically neutral to take part in an interphase partition?

(b) What is the condition of equilibrium for a species involved in interphase partition?

4.7.7. Hexanoic acid and 1-aminohexane, adjusted to pH 12 with NaOH, was passed through a cation-exchange column loaded with NaOH at pH 12. State the principal species that will be eluted and the order in which they are expected for the following $CH_3CH_2CH_2CH_2CH_2CO_2H$, (hexanoic acid), and $CH_3CH_2CH_2CH_2CH_2CH_2NH_2$ (1-aminohexane).

4.7.8. At your convenience, look up the pK_a values for trimethylamine, dimethylamine, methylamine, and ammonia. Predict the order of elution of these compounds from a cation-exchange column eluted with a gradient of increasing pH, beginning at pH 7.

4.7.9. Find the voltage of a cell, the right half-cell of which has 0.50 M KCl (aq) with AgCl (s) electrode and the left half-cell contains 0.01 M $Cd(NO_3)_2$ (aq). Write the net cell reaction and state whether it is spontaneous in the forward or reverse direction. Given the following:

right half-cell: $2\,AgCl\,(s) \; + \; 2e^- \; \rightleftharpoons \; 2\,Ag\,(s) \; + \; 2\,Cl^- \qquad E^0_+ = 0.222\ V$

left half-cell: $Cd^{2+} \; + \; 2\,e^- \rightleftharpoons Cd\,(s) \qquad\qquad\qquad E^0_- = 0.402\ V$

4.7.10. (a) Provide the difference between the following: electric charge (q, coulombs), electric current (I, ampares), and electric potential (E, voltage). (b) How many electrons are in 1 coulomb? (c) How many coulombs are in 1 mole of charge?

4.7.11. (a) How many coulombs of charge are passed from reductant to oxidant when 1.00 g of Te is deposited? (b) If Te is formed at the rate of 1.00 g/h, determine how much the current is flowing.

4.7.12. Suppose you want to extract I_2 from 100 mL of aqueous phase solution. The partition coefficient constant of I_2 for extraction with CCl_4 is 85. (a) What fraction of I_2 remains in the aqueous phase after a single extraction with 100 mL of CCl_4. (b) What fraction of I2 remains in the aqueous phase after 10 extractions with10 mL of CCl_4 (total CCl_4 volume of 100 mL)? (c) What will be the minimum volume of CCl_4 required to achieve 99.9% extraction of the I_2 upon using five extractions?

4.8 References

Bhattacharyya, L. and Rohrer, J. S. (2012). *Applications of Ion Chromatography for Pharmaceutical and Biological Products*. New Jersey: John Wiley & Sons.

Fritz, J. S. and Gjerde, D. T. (2009). *Ion Chromatography* 4th ed., Weinhein: Wiley-VCH Verlag GmbH & KGoA Weinhein.

Galus, Z. (1994). *Fundamentals of Environmental of Electrochemistry Analysis,* Ellis Horwood, New York.

Hamnett, A., Hamann, C. H. and Vielstich, W. (1998). *Electrochemistry,* Wiley, New York.

Harris, T. M. (1995). Revitalizing the Gravimetric Determination in Quantitative Analysis," *J. Chem. Ed. 72,* 355.

Ishihara, T., Kadoya, T. and Yamamoto, S. (2007). Application of a Chromatography Model with Linear Gradient Elution Experimental Data to the Rapid Scale-up in Ion-exchange Process Chromatographyof Proteins. *J. of Chrom. A.* 1162, 34.

Mochalski, P. Wzorek, B. Sliwka, I. and Amannm, A. (2009). Chromatogram B *Analyt Technol. Biomed. Life Sci. 877,* 1856.

Rajeshwar, K. and Ibanez, J. G. (1997). *Environmental Electrochemistry,* Academic Press, San Diego.

Reys, D. R., Iossifidis, D., Auroux, P. A. and Manz. A. (2002). Micro Total Analysis Systems: 1. Introduction, Theory, and Technology" *Anal. Chem. 74,* 2623.

Selco, J. I., Roberts Jr., J. L. and Wacks, D. B. (2003). "The Analysis of Seawater: A Laboratory-Centered Learning Project in General Chemistry," *J. Chem. Ed. 80,* 54.

Smith, M. J. and Vincent, C. A. (2002). Popular Science Information- Electrochemistry, *J. Chem. Ed. 79,* 851.

Storer, D. A. and Sarquis, A. M. (2000). "Measuring Soil Phosphates Using-Ion-Exchange Reseign", *J. Chem. Edu. 77.* 748.

Thompson, R. Q. and Ghadaali, M. (1993). Microwave Drying of Precipitates from Gravimetric Analysis," *J. Chem. Ed. 70,* 170.

Walton, S. K., Abney, M. B. and Levan, M. D. (2006). CO_2 Adsorption in Y and X Zeolites Modified by Alkali metal Cation-exchange, *Microporous and Mesoporous Materials, 91,* 78.

Williams, J. P., West, K. J. and Erickson, K. L. (1992). "Separation of Aspirin from Acetaminophen and Caffeine in an Over-the-Counter Analgesic Tablet: A Solid-Phase Extraction Method," *J. Chem. Ed. 69,* 669.

Chapter 5

QUANTITATIVE CHEMICAL ANALYSIS

5.1   Introduction

The process of determining the amount of a particular substance in a sample is called *quantitative analysis.* As is always the case, the substances of interest are in a matrix that includes many other unwanted substances. Often, the substance of interest is present in the sample in only minute or trace amounts. The challenge for the analytical chemist is how to deal with only the analyte of interest and to determine the quantities of just those analytes without being influenced by the nature and amount of other materials present. Quantitative chemical analysis deals with the determination of amounts (relative amounts expressed in a percentage). Quantitative chemical analysis is indispensable in many branches of science in the earth sciences, in biology and industry, as well as in the analysis of raw materials and finished products including clinical medicine, agriculture and other fields.

5.2   Units Involved in the Expression of Amount

The expression of amount of material can be looked at in different forms. Chemically, the number of moles of elements or compounds in the sample is a very useful unit. Despite the fact that we have seen solutions and concentration units in chapter 2, section 2.3 it is still useful to look at general units used in the determination of amount. The number of moles is the ratio of the mass in grams and *formula weight* (FM). The term *molecular weight* (MW) is used for the molecular species. The existing relationship between mass and FW or MW is valid for pure materials of known chemical formula. A *mole* of substance is Avogadro's number (6.022×10^{23}) of the smallest unit (atoms, molecules, ions, etc.) that retain the qualities of the substance. The term mole is exceptionally important in chemical analysis because the recognition of the analyte often takes place at the atomic and molecular level.

When analyte is dissolved in appropriate liquid (frequently referred to us *solvent*), it becomes uniformly dispersed in the solvent. The amount of analyte (or *solute*) present in a given amount of solution is generally called *concentration*. Several ways of expressing the concentration of dissolved species including *molarity* and *molality,* which have been discussed in detail in chapter 2, section 2.3.

89

Example 5.1

Calculate the molarity of $CaCO_3$ in a solution prepared by placing 0.5471 g $CaCO_3$ in a 1000 mL volumetric flask and diluting to the mark with distilled water, H_2O. The formula weight of $CaCO_3$ is 100.09 g/mol. Assume the calcium carbonate dissolves completely.

Solution

Since molarity is moles per litre, it is convenient to convert the grams of $CaCO_3$ to moles. The molarity is then calculated by dividing moles of $CaCO_3$ by a litre of solution.

$$= 0.5471 \text{ g} \left(\frac{1 \text{ mol}}{100.09 \text{ mL}}\right)\left(\frac{1}{100.0 \text{ mL}}\right)\left(\frac{1000 \text{ } mL}{1 \text{ } L}\right)$$
$$= 0.0547 \text{ M } CaCO_3$$

Practice 5.1

(a) Consider a piece of iron which is contains 31.83 mol of iron atoms. How much does it weigh? MW for iron is 55. 86 g/mol.
(b) Calculate the number of moles of NaCl in 50.0 mL of a 0.200 M NaCl solution.
Solution (a) 1.78×10^3 g (b) 0.010 mol NaCl.

5.3   Methods of Quantitative Analysis

This is always concerned with the relative amounts of constituents in a sample. The relative amounts, expressed as weight percentages, are obtained in a known amount of sample. A method of quantitative analysis involves the measurement of some physical properties such as mass. However, in more significant sense a distinction can be made between physical (including physical chemical), and chemical methods.

5.3.1 Chemical Methods

This involves application of chemical reaction on the constituents being determined. The approach differs from others by being ideally based on a stoichiometric reaction.

whereas C stands for constituent, and R for the chemical reagent. One can determine the amount x of C from the amount, w, of the product C_mR_n or when the amount of the reaction product is found by weight we speak gravimetric analysis. When amount of constituent is found through reaction with reagent, we speak of *titrimetric analysis*. When the quantity of reagent required is obtained after measuring the volume of standard solution, we speak of *volumetric analysis*.

Indeed, gravimetric and volumetric methods were the first to be used in quantitative analysis. For the case when the product is in a gaseous state, and can be determined from the volume of gas at known temperature and pressure, we speak of *gasometric methods*.

5.3.3   Physical and Physicochemical Methods.

This is a method which is plainly physical and no reaction is involved in any determination. There are many physical properties which are functions of either of the mass of a substance or its concentration. For example, the colour intensity of a solution of permanganate is proportional to be concentration of MnO_4^-. Thus, by measuring colour intensity we can determine concentration of manganese. Often, the manganese would be present originally in the divalent state and oxidize to permanganate. This is a physical chemical method.

With some exceptions, physical and physicochemical are instrumental methods, and they require the use of instruments other than balance and burette. If chemical reaction is involved, it does not require being stoichiometric. Physical methods are concerned with energy and chemical methods are concerned with mass. The most important physical and physicochemical methods are based on emission and absorption of radiation.

5.4   Gravimetric Analysis

In principle, when analyte has been separated from the sample mixture by precipitation, the amount of it can be determined by weighing. This method is one of classic techniques called gravimetric analysis. Gravimetric analysis has been largely replaced by instrumental methods of analysis, which are faster and less labour intensive. Nevertheless, gravimetric analysis has remained among the most accurate method of analysis when such determination is carried out by a skilled analyst. This method works most, if not for all, inorganic anions and cations, including neutral species such as carbon dioxide, iodine, sulfur dioxide and water. A number of organic substances can also be determined gravimetrically. These include cholesterol in cereals, nicotine in pesticides, lactose in milk products, and aspirin.

When using gravimetric analysis we will not be weighing the analyte as it was in the sample. It has now reacted with the precipitating reagent to create a new species. The number of molecules of the new species will be exactly related to the number of atoms of the analyte if the formation of the precipitate is exactly stoichiometric. Suppose Ba^{2+} ion is separated from a mixture by precipitation with SO_4^{2-}, as $BaSO_4$ precipitate. The number of moles of $BaSO_4$ is equal to the number of Ba^{2+} in the original sample if the ratio of Ba^{2+} to SO_4^{2-} in the $BaSO_4$ is exactly 1:1. As we have seen in chapter 2, section 2.1 regarding stoichiometry, while some chemical reactions have stoichiometric factors such as integer number, other compounds have no exactly stoichiometric. Because of the latter challenges we must then be able to determine the number of moles of precipitate by weighing it. This will be possible upon doing a thorough removal of the rinse solvent and achievement of a stable form of the precipitate.

Quantitative transfers of the precipitated material to a weighing vessel often present challenges. These include small particles which remain in the pores of the filter, which create a large error. Thus, the precipitate should be filtered in such a way that it can be easily dried and weighed in a filter medium. This is always possible if a glass or ceramic filter container is used. The filter container used to dry precipitate is called a *crucible*. The crucible is dried and weighed before the filtration. It is advised that once the crucible is weighed, tongs or gloves must be used to handle it to avoid adding the weight of fingerprints.

A major challenge in the determination of precipitate amount through weighing is on how to obtain a stable composition of the precipitate. It is difficult to bring most of the materials to a stable and reproducible composition. When the temperature of the crucible is raised to dry the precipitate, the precipitate may begin to lose some water of hydration. Before the latter is gone, the material may undergo further decomposition. A good example is carbonate losing CO_2 to form oxides, and hydroxides losing H_2O to form oxides. Some materials have a reasonable range of temperatures over which the composition is stable, and suitable product for gravimetric analysis can be obtained.

For the purpose of calculation of the amount of analyte from the weight of the material in the crucible, one should know exactly the ratio of the weight of the substance under consideration and the substance weighed. Suppose Fe^{3+} ion is precipitated with OH^- ion, the precipitate formed is $Fe(OH)_3$. When it is weighed, it is in the form of Fe_2O_3. Thus, the mass of the iron in x g of Fe_2O_3 can be calculated as in expression 5.1 below:

$$grams\ of\ Fe = x\ \frac{MW_{Fe}}{\frac{1}{2}MW_{Fe_2O_3}} = x\ \frac{55.847}{79.845} = 0.699\ x \qquad (5.1)$$

In equation 5.1 a half of the molecular weight of Fe_2O_3 is used in the denominator because each mole of Fe_2O_3 contains two moles of Fe. The factor that relates the weight of the Fe_2O_3 to the weight of the iron (0.699) is called the *gravimetric factor*.

Such type of analysis is good for students because the procedure demands students to provide accurate and precise results.

Simple gravimetric analysis that you will often encounter is the determination of Cl- by precipitation with Ag+:

$$Ag^+(aq) \quad + \quad Cl^-(aq) \longrightarrow AgCl(s)$$

The mass of AgCl product tells us how many moles of AgCl were produced. For every mole of AgCl, there must have been 1 mole of Cl- in the unknown solution.

Example 5.2:

A 10.00 mL solution containing Cl- was treated with excess $AgNO_3$ to precipitate 0.4368 g of Ag Cl. What was the molarity of Cl- in the unknown?

Solution

Formula mass of AgCl is 143.321 g mol-1. A precipitate was weighed 0.4368 g

$$\frac{0.4368 \ g \ AgCl}{143.321 \ g \ AgCl / mol \ AgCl} = 3.048 \times 10^{-3} \ mole \ AgCl$$

Because 1 mole of AgCl contains 1 mole of Cl-, there must have been 3.048 x 10-3 mol of Cl- in the unknown.

$$[Cl^-] = \frac{3.048 \times 10^{-3} \ mol}{0.0100 \ L}$$

$$[Cl^-] = 0.3048 \ M$$

One has to be cautious in carrying out gravimetric analysis because conditions must be controlled to selectively precipitate only the specie of interest. Potentially interfering substances may need to be removed by filtration process prior to further analysis.

Example 5.3

To determine the amount of Cl- in 30.00 mL of an unknown solution, an excess of $AgNO_3$ is added. The AgCl precipitate was found to weigh 0.3912 g. What is the concentration of Cl- parts per thousand (ppt) in the original solution? Assume the density of original solution is 1.00 g/mL.

Solution:

$$Ag^+(aq) \quad + \quad Cl^-(aq) \rightleftharpoons AgCl(s)$$

The amount of Cl- in the precipitate is

$$= (0.3912\ g\ AgCl)\left(\frac{1\ mol\ AgCl}{143.321\ g}\right)\left(\frac{1\ mol\ Cl^-}{1\ mol\ AgCl}\right)$$

$$x\left(\frac{35.4527\ g\ Cl^-}{mol\ Cl^-}\right) = 0.09677g\ Cl^-$$

From the fact that the density of the solution is about 1 g/mL, the amount in parts per thousand is calculated from the milligrams solute per litre of solution.

$$\frac{96.77\ mg\ Cl^-}{30.00\ mL} = 0.09677g\ Cl^-$$

5.5   Precipitation

The ideal product of gravimetric analysis always should be insoluble, easily filterable, and very pure, and should possess a known composition. Although few substances meet all these requirements, appropriate techniques are helpful in optimizing the properties of gravimetric precipitates. For example, the solubility of a precipitate is usually decreased by cooling and a much purer form of the solid will recrystallize from the solution. Particles of the precipitate should not be so small that they clog or pass through the filter. Large crystals are the product of slow cooling; also they have less surface area to which foreign species may become attached within the crystal lattice. Precipitation conditions have much to do with the resulting particle size.

5.5.1   Desired Properties of a Precipitate

The desired ion or molecule in gravimetric method is isolated by the addition of appropriate reagent which interacts to form an insoluble or slightly soluble compound. Once the precipitate is formed, it is filtered, washed, dried and either can be weighed directly or transformed to another compound which is more convenient to weigh.

For the purpose of good analytical results, the precipitate should be insoluble, easily filterable, free of impurities, and have a well defined chemical composition. If there is any doubt regarding the composition variations, it must be easily convertible into a compound which possesses a known composition. A good example and one easy to remember is hydrated ferric oxide, $Fe_2O_3.nH_2O$, usually obtained when ammonia is added to ferric salt solution, including a variable amount of water. However, upon heating the hydrated ferric oxide is converted to the anhydrous ferric oxide, Fe_2O_3.

5.5.2   The Crystallization of Precipitate

The crystallization process is dynamic. This proceeds with substantial tension between the tendency to crystallize and the tendency to dissolve. Thus, consider a situation where M and X entities are leaving the surface of the crystal and other M and X entities in the solution are continuing to form crystal lattice. During crystallization there are two effects which influence the composition of the product and its analytical usefulness.

(i) Occlusion often happens when a second species, for example N, is present in solution, whereby to some extent N resembles M (or X) in charge, conformation and size. Though the reaction between N and M does not produce an insoluble product, there is a chance for an N to temporarily take the place of an M on the crystal surface and then, under conditions of rapid growth, to become trapped there. A good example is the precipitation of Ba^{2+} and SO_4^{2-} ions. In addition, if at all some Ca^{2+} ions exist in solution, the final $BaSO_4$ precipitate will have a small portion of $CaSO_4$ impurities. The latter affect the quantitation because the weight of the precipitate formed was not limited by the amount of Ba^{2+} ions present and due to the amount of SO_4^{2-} required to form the precipitate exceeding the stoichiometric amount required by Ba^{2+} ions. Occlusion can be relatively lowered through avoiding rapid precipitate growth (by slow addition of dilute reagent and rigorous stirring) and *aging* the precipitate before separation. Often, aging takes time and the dynamic nature of the crystallization process to improve the purity of the precipitate.

5.5.3   Colloidal Suspensions

Colloidal particles have a dimension/diameter which ranges between 1-100 nm and because of such small size they are not often retained on ordinary filters. Moreover, colloidal particles are often formed as a result of excess electrical charge, and in many cases almost all colloidal particles carry charges of the same charge. A good example is when silver ion is precipitated by addition of a soluble chloride: the resulting solution contains an excess of chloride ions, which tend to be adsorbed on the surface of silver chloride particle (Figure 5.1). This means the silver chloride colloidal particles will be carrying a negative charge. The existence of such conditions promotes electrostatic repulsion between the colloidal particles, which prevents them from coalescing. Once precipitation is carried out by the addition of silver nitrate to a chloride solution, there will be an excess of silver ions present on completion of precipitation, and the silver particles are carrying a positive charge: see Figure 5.1. In addition, Brownian motion prevents their settling from the solution under the influence of gravitational forces. Furthermore, an individual particle of most colloids can be forced to coagulate and form a more filterable, non-crystalline, which settles from the solution. Coagulation of the colloids will happen when there are relatively fewer repulsive forces between the primary adsorbed ions on the surface and their counter ions in the solution. Obviously, huge repulsive forces usually offset the normal cohesive forces that exist between small particles having the same chemical composition.

5.5.4 Crystal Growth

Crystallization occurs essentially in two phases:

(i)   Nucleation, and

(ii)   Particle growth

In many cases the question of how the first crystals of precipitate form (nucleate) remains interesting for chemists. During nucleation, molecules in solution come together randomly and form small aggregates. Indeed, aggregates of a small number of species are called *clusters*. The question is how the latter are formed and at what point does a cluster begin to act more like a crystal. What is known from experiment it is that, as the crystal size decreases below 10^{-3} mm in diameter, the solubility of crystals tends to increase significantly.

Figure 5.1: Adsorbed Ions on a Colloidal Particle of AgCl

Particle growth involves the addition of more molecules to a nucleus to form a crystal. The high concentration of precipitant in the region where the reagent is added to the solution can provide excess and create the conditions required for nucleation. When a solution contains more solute than should be present at equilibrium, the solution is said to be *supersaturated* as we have seen in chapter three. Nucleation proceeds faster than particles in a highly supersaturated solution. This results in suspension of tiny particles, or worse, a colloid. In a less supersaturated solution, nucleation is slower, and nuclei have a chance to grow into larger, more tractable particles. Under these circumstances deposition of solid on the particles already formed may occur without further nucleation.

The following are among the techniques or measures to be taken to reduce supersaturation and promote the formation of large particles, which promote particle growth:

(i)   Keeping the volume of solution large so that the concentrations of analyte and precipitate are low during the period in which the precipitate is formed.

(ii)   Raising the temperature to increase solubility and thereby decrease supersaturation.

(iii)   Addition of precipitant at a slow rate with vigorous mixing, to avoid a local, highly supersaturated condition, where the stream of precipitant first enters the analyte, and

(iv)   Using homogeneous precipitation.

In precipitation separation, it is essential to reduce the amount of a precipitate that is in the form of very small crystals. Through the avoidance of colloidal particle formation one can alleviate the chance of small crystal formation. The smaller crystals which have higher solubility will dissolve in the solution that is in equilibrium with the larger crystals. The chance of nucleation under that circumstance is unlikely, so the material that dissolves from the small crystals will add to the size of the larger ones already formed. In most cases, equilibrium concentration is reached in less than a minute. In fact, crystal stabilization can take several hours, days, or even months.

5.5.5   Homogeneous Precipitation

It is apparent that nucleation occurs in the regions of temporary excess precipitant that results from drop-wise addition of the reagent. Nucleation can be avoided by uniformly increasing the reagent concentration throughout the solution. This procedure is called homogeneous precipitation. A good example is where the precipitant is generated slowly by a chemical reaction of urea which decomposes in boiling water to produce OH^- (expression 5.2) and the use of sulfamic acid to homogeneously generate SO_4^{2-} (see expression 5.3).

$$\underset{\substack{H_2N \qquad NH_2}}{\overset{\overset{\displaystyle O}{\parallel}}{C}} \;+\; 3\,H_2O \;\xrightarrow{\text{Heat}}\; CO_2 \;+\; 2\,NH_4^+ \;+\; 2\,OH^- \qquad (5.2)$$

The above expression implies that the pH of a solution can be raised very gradually. Slow OH^- formation enhances the particle size of Fe (III) formate $(Fe(CO_2)_3.xH_2O)$ precipitates.

$$H_3NSO_3 \;+\; 2\,H_2O \;\longrightarrow\; NH_4^+ \;+\; H_3O^+ \;+\; SO_4^{2-} \qquad (5.3)$$

Sulfamic acid is relatively unstable in water and slowly undergoes the reaction, as can be seen in expression 5.3. Upon heating the solution to promote the reaction, the sulfate ion is slowly released throughout the solution. The technique is most useful in precipitating Ca^{2+}, Ba^{2+}, Sr^{2+}, or Pb^{2+}. Homogeneous precipitation is advantageous due to the lack of local excess concentrations, which decreases the generation of small crystals and slow crystal growth, which in turn reduces occlusion.

5.5.6   Precipitation in the Presence of Electrolyte

Ionic compounds are often precipitated in the presence of an electrolyte. This can be easily comprehended by using the case of AgCl, which is commonly formed in 0.1 M HNO_3 (Figure 5.2). A good example is a colloidal particle of AgCl growing in a solution containing excess Ag^+, H^+ and NO_3^-. The particle has excess net positive charge because of adsorbed (to be attached to the surface) Ag^+ ions on exposed chloride ions. It is good to remember that absorption involves penetration beyond the surface to the inside. The region of solution surrounding the particle is called the ionic atmosphere (Figure 5.2). It has a net negative charge, because the positively charged particle attracts anions and repels cations from ionic atmosphere. Colloidal particles must collide with one another to coalesce. Nevertheless, the negatively charged ionic atmospheres of the particles repel one another.

Figure 5.2: Colloidal Particle of AgCl Growing in Solution of Excess Ag^+, H^+ and NO_3^-

The particles, therefore, must have enough kinetic energy to overcome electrostatic repulsion before coalescence. It is apparent that heat promotes coalescence by increasing the particle kinetic energy. Increasing electrolyte concentration (HNO_3 for AgCl) decreases the volume of the ionic atmosphere and allows particles to come closer together before electrostatic repulsion gets or becomes significant. Because of the latter reason, most gravimetric precipitations are done in the presence of an electrolyte. Often, the addition of a suitable electrolyte on charged colloidal particles will help to neutralize it.

This promotes flocculation of the precipitate. The suitability of electrolyte and the concentration necessary to produce flocculation depend on the nature and the charge of the colloidal particles. Generally speaking, negatively charged particles are most efficiently flocculated by salts with polyvalent cations, whereas particles with positive charges respond better to electrolytes with polyvalent anions. Good choice of electrolyte should always be made to avoid electrolyte which can interfere with the precipitation reaction.

Once a colloidal dispersion is flocculated by the addition of an electrolyte, such a process can be reversed by the removal of the electrolyte. This usually can be removed during washing of the filtered precipitate; the concentration of the coagulating ions drops to the extent that they no longer neutralize the charge on the particles, and there is a possibility of the precipitate going back into colloidal dispersion. The latter phenomenon is called *peptization*. This is often avoided by washing the precipitate with easily removable electrolyte in subsequent operations. Thus, for example, silver chloride precipitate is often washed with a dilute nitric acid, because nitric acid adhered to the precipitate evaporates during the drying process and leaving pure silver chloride. Furthermore, ammonium nitrate can be used in washing precipitates if a particular precipitate is to be calcined at temperature above 500 ºC. Obviously, at such temperature ammonium nitrate is voltalized and contamination of the precipitate is avoided.

5.6   Digestion
The time frame at which the precipitates stand in the presence of the hot mother liquor is called digestion. This is a slow recrystallization of the precipitate. The particles' size increases and impurities tend to be expelled from the crystal surface. Digestion is often done in hot mother liquor for the purpose of promoting particle growth and recrystallization.

5.7   Purity
Adsorbed impurities are often bound to the surface of a crystal. Adsorbed impurities within the crystal are classified as inclusion or occlusions. Inclusions are impurity ions that randomly occupy sites in the crystal lattice normally occupied by ions that belong to the crystal. *Inclusions* are more likely when the impurity ion has a size and charge similar to one of the ions that belongs to the products.

Occlusions are pockets of impurities that are literally trapped inside the growing crystal.

Co-precipitation tends to be worst in colloidal precipitation (which have a large surface area), such as $BaSO_4$, $Al(OH)_3$, and $Fe(OH)_3$. Many procedures require washing away the mother liquor, redissolving the precipitate and reprecipitating the product. For second precipitation, the concentration of co-precipitation becomes lower than the first precipitation; the extent of co-precipitation tends to be lower. Moreover, washing the precipites on a filter helps to remove droplets of liquid containing excess solute. Some can be washed with water, but many require electrolyte to maintain coherence. However, the electrolyte used for washing must be volatile, so that it can easily be lost during drying process. Some examples of volatile electrolytes include HNO_3, HCl, NH_4Cl and $(NH_4)_2CO_3$. In fact, some precipitates are washed with a volatile electrolyte in order to avoid peptization.

5.8   Product Composition

Final product should always have a known and stable composition. A *hygroscopic* substance is one that picks up water from the air, and is therefore difficult to weigh accurately. Many precipitates contain a variable quantity of water, and must be dried under conditions that provide a known (possibly zero) stoichiometry of water. Ignition (strong heating) is used to change the chemical form of some precipitates. For example, igniting $Fe(HCO_2)_2 \cdot nH_2O$ at 850 °C for 1 hour gives Fe_2O_3. The final product must always be heated to dryness or ignited to achieve a reproducible, stable composition.

Example 5.4

Relating mass to product to mass of reactant; the piperazine content of an impure commercial material can be determined by precipitating and weighing diacetate.

$$\text{Piperazine} + 2\,CH_3CO_2H \longrightarrow \text{Piperazine diacetate}$$

Piperazine	Acetic acid	Piperazine diacetate
FM 86.136	FM 60.052	FM 206.240

In one experiment, 0.313 g of the sample was dissolved in 25 mL of acetone, and 1 mL of acetic acid was added. After 3 minutes the precipitate was filtered, washed with acetone, dried at 110 °C, and found to weigh 0.7121 g. What is the weight percent of piperazine in the commercial material?

Solution:

For each mole of piperazine in the impure material of 1 mol of product is formed.

$$Mole\ of\ product = \frac{0.712\ g}{206.24\ g.mol^{-1}} = 3.45\ x\ 10^{-3} mol$$

This many moles of piperazine correspond to

Grams of piperazine = $(3.453\ x\ 10^{-3}\ mol)\ x\ (86.136\ g.\ mol^{-1})$

Grams of piperazine = 0.297 g, this gives a percentage of pepirazine in analyte as:

$$\frac{0.297\ g}{0.313\ g}\ x\ 100 = 94.89\%$$

The above problem can alternatively be worked out as follows:

Consider 206.240 g (1 mol) of product will be formed for every 86.136 g (1 mol) of pepirazine analyzed. Because 0.712 g of product was formed, the amount of reactant is given by:

$$\frac{x\ g\ piperazine}{0.712\ product} = \frac{86.136\ g\ piperazine}{206.243\ g\ product}$$

$$\frac{86.136\ g\ piperazine\ x\ 0.712\ product}{206.243\ g\ product}$$

$$x = 0.297\ g\ piperazine$$

The quantity 86.136 g/206.240 g is the *gravimetric factor* relating the mass of starting material to the mass of product. This is done based on the assumption that if there were impurities in pepirazine they did not precipitate, otherwise the precipitate of the product would be high.

Example 5.5

Solid residue weighing 8.4448 g from the aluminum refining process was dissolved in acid, treated with 8-hydroxyquinoline, and ignited to give Al_2O_3 weighing 0.8554 g. Find the weight percent of Al in the original mixture.

$$Al^{3+} \longrightarrow \left(\begin{array}{c} \\ \end{array} \right)_3 \overset{heat}{\longrightarrow} Al_2O_3$$

FM 26.982 g/mol

0.8554g
FM 101.961 g/mol

Solution:
Each mole of product (Al_2O_3) contains two mole of Al. The mass of product tells us the mole of product and from this one can determine the mole of Al. The mole of product are $(0.8554\ g)/(101.961\ g/mol) = 0.008389\ mol\ Al_2O_3$.

$$Mole\ of\ Al\ in\ unknown = \frac{2\ mol\ Al}{mol\ Al_2O_3} \times 0.008389\ mol\ Al_2O_3 = 0.01677\ mol\ Al$$

The mass of Al is $(0.01677_9\ mol)(26.982\ g/mol) = 0.4527_3\ g$ Al. The weight percentage of Al in the unknown can be calculated as follows:

$$wt\%\ Al = \frac{0.4527_3\ g\ Al}{8.4448\ g\ unknown} \times 100 = 5.361\%$$

Example 5.6
The element cerium, discovered in 1839, and named after the asteroid Ceres, is a major component of flint lighters. To find the Ce^{4+} content of a solid, an analyst dissolved 4.37 g of the solid and treated it with excess iodate to precipitate $Ce(IO_3)_4$. The precipitate was collected, washed, dried and ignited to produce 0.104 g of CeO_2.

(a) How much cerium is contained in 0.104 g of CeO_2?
(b) What was the weight percentage of Ce in the original sample?

Solution:
(a)

$$\frac{0.104\ g\ CeO_2}{172.115\ g\ CeO_{2/mol}} = 6.043 \times 10^{-4}\ mol\ CeO_2$$

$$\frac{1\ mol\ Ce}{1\ mol\ CeO_2} \times 6.043 \times 10^{-4}\ mol\ CeO_2 = 6.043 \times 10^{-4}\ mol\ Ce$$

$$(6.043 \times 10^{-4} \, mol \, Ce)\left(140.116 \, g \, \frac{Ce}{mol} Ce\right) = 0.08466 \, g \, Ce$$

$$(b) \; wt\% \; Ce = \frac{0.08466 \, g \, Ce}{4.37 \, g \, uknown} \times 100 = 1.94\%$$

Example 5.7

Consider a mixture of two solids, $BaCl_2.2H_2O$ (FM 244.26 g/mol) and KCl (74.55 g/mol), in unknown ratio. (The notation $BaCl_2.2H_2O$ means that a crystal is formed with two water molecules for each $BaCl_2$.) Upon heating unknown to 160 °C for 1 h, the water of crystallization is driven off, as can be seen below:
A sample originally weighing 1.784 g weighed 1.562 g after heating. Calculate the weight percent of Ba, K, and Cl in the original sample.

Solution:

Formula and atomic masses: Ba (137.327), Cl (35.453), K (39.098), H_2O (18.015), KCl (74.551 g/mol), $BaCl_2.2H_2O$ (244.26).H_2O weight lost = 1.784 g – 1.562 g = 0.222 g which is equal to 1.230×10^{-3} mol of H_2O. For every two moles of H_2O lost one mole of $BaCl_2$. $2H_2O$ must have been present. 1.230×10^{-2} mol of H_2O implies that 6.150×10^{-3} mol of $BaCl_2.2H_2O$ must have been present. This much $BaCl_2.2H_2O$ equals 1.502 g. The Ba, and Cl contents of the $BaCl_2.2 \, H_2O$ are as follows:

$$Ba = \frac{137.33}{244.26} (1.502) = 0.885 \, g$$

$$Cl = \left(\frac{2\,(35.453)}{244.26}\right) (1.502) = 0.436 \, g$$

Due to the fact that the total sample weighs 1.784 g and contains 1.502 g of $BaCl_2.2 \, H_2O$, the sample must contain 1.784 g – 1.502 g = 0.282 g of KCl which contains the following:

$$K = \left(\frac{39.098}{74.551}\right) (0.282) = 0.14\,8 \, g$$

$$Cl = \left(\frac{35.453}{74.551}\right) (0.282) = 0.134 \, g$$

Weight percentage of each element is calculated as follows:

$$Ba = \frac{0.845}{1.784} \times 100 = 47.35\%$$

$$K = \frac{0.148}{1.784} \times 100 = 8.28\%$$

$$Cl = \frac{0.436 + 0.134}{1.784} \; x \; 100 = 31.95\%$$

5.9 Problems

5.9.1 (a) Explain clearly the difference between adsorption and absorption.
(b) Make a clear distinction between an occlusion from an inclusion.
(c) How can you determine the number of moles in a given mass of a substance?

5.9.2 (a) What are desirable properties of gravimetric precipitate?
(b) Why is relative high supersaturated undesirable in a gravimetric precipitation?
(c) What does it mean if a solution is said to be 0.40 molar in a particular substance?

5.9.3. Why reprecipitation would be employed in a gravimetric analysis?

5.9.4. (a) Why are most of ionic precipitates washed with electrolyte solution instead of pure water?
(b) Why is it not a good idea to wash AgCl precipitate with aqueous $NaNO_3$ rather than with HNO_3 solution?

5.9.5. What measures can be taken to decrease supersaturation during precipitation?

5.9.6. A solution containing 1.263 g of unknown potassium compound was dissolved in water and treated with excess sodium tetraphenylborate, $Na^+B(C_6H_5)_4^-$ solution to precipitate 1.003 g of insoluble $K^+B(C_6H_5)_4^-$ (FM 358.33 g/mol). Find the wt% of K in the unknown.

5.9.7. A 50.00 mL solution containing NaBr was treated with excess $AgNO_3$ to precipitate 0.2146 g of AgBr (FM 187.772 g/mol). What was the molarity of NaBr in the solution?

5.9.8. Finely ground mineral of 0.6324 g was dissolved in 25 mL of boiling 4 M HCl and diluted with 175 mL H_2O containing two drops of methyl red indicator. The solution was heated to 100 °C and 50 mL of warm solution containing $(NH_4)_2C_2O_4$ were slowly added to precipitate CaC_2O_4. Then 6 M NH_3 was added until the indicator changed from red to yellow, showing that the liquid was neutral or slightly basic. After slow cooling for 1 h, the liquid was decanted and the solid transferred to a filter crucible and washed with cold 0.1 wt% $(NH_4)_2C_2O_4$ solution five times until no Cl^- was detected on addition of $AgNO_3$ solution. The crucible was dried at 110 °C for 1 h and then at 500 ± 25 °C in a furnace for 2 h.

The mass of empty crucible was 18.2311 g and the mass of the crucible with $CaCO_3$ (s) was 18.58467 g.

(a) Find the wt% of Ca in the mineral.

(b) Why is the unknown solution heated to boiling and the precipitant solution, $(NH_4)_2C_2O_4$ also heated before slowly mixing the two?

(c) What is the purpose of washing the precipitate with 0.1 wt% $(NH_4)_2C_2O_4$?

(d) What is the purpose of testing the filtrate with $AgNO_3$ solution?

5.9.9. A mixture of Al_2O_3 (s) and CuO (s) weighing 18.371 mg was heated under H_2 (g) in a thermogravimetric experiment. Upon reaching a temperature of 1000 °C, the mass was 17.462 mg and the final products were Al_2O_3 (s), Cu (s) and H_2O (g). Find the weight percentage of Al_2O_3 in the original solid mixture.

5.9.10. Students of analytical chemistry at SEKOMU varied the procedure in the gravimetric analysis of Ba^{2+} by precipitation of $BaSO_4$. The following sequence was carried out twice, giving 0.495 and 0.499 g $BaSO_4$.

(i) Pipette 50 mL of unknown $BaCl_2$ solution (~0.05 M) into a 400 mL beaker and add 100 mL of H_2O.

(ii) Add 1 mL of 12 M HCl to increase solubility of $BaSO_4$ and decrease supersaturation during precipitation. Solubility is increased because the

reaction   consumes SO_4^{2-}.

(iii) Heat to 90 °C to increase the solubility of $BaSO_4$ and promote crystallization during precipitation.

(iv) Add 50 mL of 0.15 M Na_2SO_4 all at once.

(v) Digest the precipitate at 90 °C for 1 h in the mother liquor.

(vi) Decant the mother liquor (pour it off), wash the precipitate with hot water, and decant the wash water.

(vii) Collect the precipitate by filtration through ashless filter paper and wash it with hot water until no Cl^- is detected in the filtrate (by precipitation with Ag^+).

(viii) Transfer the paper and precipitate to a porcelain crucible and ignite it over a Bunsen burner to burn away the paper, and leave pure $BaSO_4$.

(ix) Cool the crucible in a disiccator and weigh it. Re-heat and reweigh until successive weights agree within ± 0.3 mg.

(a) When the initial solution contained 0.5 mmol $CaCl_2$ in addition to $BaCl_2$, 0.503 g and 0.507 g of the products was obtained. What is the possible explanation for this? (b) The procedure in (a) was carried out with digestion for 20 h, instead of 1 h, at 90 $^\circ$C to give 0.496 and 0.499 of product. Provide a possible explanation.

5.9.11. How many moles are there in 100.0 mL of mercury at 25 $^\circ$C? (The density of mercury at 25 $^\circ$ C is 13.59 g/mL and its atomic weight is 200.59 g/mol).

5.9.12. Why is a concentration of the precipitating reagents in excess of the K_{sp} product generally necessary to begin the precipitation process?

5.9.13. Determine the amount of Cl^- in 30.00 mL of an unknown solution, when an excess of $AgNO_3$ is added. The AgCl precipitate formed was found to weigh 0.3912 g. What is the concentration of Cl^- (ppt) in the original solution? Assume a density of 1.00 g/mL for the original solution.

5.10 References

Beck, C. M. (1994). "Classical Analysis: A Look at the Past, Present, and Future," *Anal. Chem. 66*, 224A.

Ghadiali, M. and Thompson, R. Q. (1993). Microwave Drying of Precipitate for Gravimetric Analysis," *J. Chem. Ed. 70*, 170.

Gupta, A. K. Lloyd, D. J. and Acont, S. (2001). Precipitationon Hardening in Al-Mg-Si, *Mat. Sc. and Eng. 316*, 11.

Harris, D. C. (2005). Exploring Chemical Analysis, 3rd ed. W. H. Freeman and Company, New York NJ.

Harris, T. M. (1995). "Revitalizing the Gravimetric Determination in Quantitative Analysis Laboratory," *J. Chem. Ed. 72*, 355.

Hill, J. O. and Magee, R. J. (1988). "Advanced Undergraduate Experiments in Thermoanalytical Chemistry," *J. Chem. Ed. 65*, 1024.

Roberts; J. L. Selco, J. I. and Wacks. D. B. (2003). The Analysis of Seawater A Laboratory-Centred Learning Project in General Chemistry." *J. Chem. Ed. 80*, 54.

Sørensen, M. A., Stackpoole, M. M., Frnkel, A. I. Bordia, B. K. Korshin, G. V. and Christensen, T. H. (2000). "Aging of Iron Hydroxides by Heat Treatment and Effects on the Heavy Metal Binding," *Environ. Sci. Technol. 34*, 3991.

Su, J. Dong, O. and Kang, B. (2005). Research on Agig Precipitation in a Cu-Cr-Zr-Mg Alloy, *Mat. Sc. and Eng. 392*, 422.

Torrades, F. and Castellvi, M. (1994). "Spectrophotometric Determination of Cl^- in $BaSO_4$ Precipitation," *Fresenius J. Anal. Chem. 349*, 734.

Chapter 6

FORMATION OF COMPLEX COMPOUNDS

6.1 Introduction

A fairly old definition states that a complex compound is one formed through the union of two or more molecules which are capable of separate existence; that is, molecules in which all the primary valences are satisfied. An example of such combinations are exceedingly numerous and may include hydrates $CuSO_4.5H_2O$, $MgCl_2.6H_2O$, ammoniates $Cu(NO_3)_2.4NH_3$; $CoCl_3.6NH3$; double salts, $PtCl_4.2KCl$, $BeF_2. 2KF$; organometallic addition compounds, $SnCl_4.2CH_3OH$, $PCl_5.2CCl_4$, etc. Nevertheless, the above definition is not a particularly satisfactory one. If we interpret it strictly, we would have salts such as ammonium chloride, sodium sulfate and hydroxide, such as ferric hydroxide, falling into the category of complex compounds, since their respective compounds can be written as $NH_3.HCl$, $Na_2O.SO_3$, and $Fe_2O_3.3H_2O$. This interpretation makes the definition too broad and consequently restricts its usefulness.

A better definition describes a complex compound as one which results from an interaction of a Lewis acid with a Lewis base. That is, the complex molecule will have one or more coordinate covalent bonds. Strictly, this places ammoniates, hydrates and double salts in the class of complex compounds, but excludes hydroxides and salts such as carbonates and sulfates.

As can be seen in the above examples, a metal ion is involved. The role of metal ion in this case is acting as coordination center. The species that bond to the coordination center in a complexation reaction are called ligands. It is common practice that several bonds are formed between the coordination center and one or more ligands. The number of ligand bonds formed around the coordination center is called coordination number.

6.2 Complex Metal Ions

The free electronic orbitals of metal ion are progressively filled with electrons furnished by an electron donor molecule or ion. In other words, the metal ion acts as an electron pair acceptor or Lewis acid, while the donor ion or molecule is a Lewis base. The definition includes reactions such as the one shown hereunder:

$$Ag^+ \; + \; 2\!:\!NH_3 \; \longrightarrow \; \left[H_3N\!:\!Ag\!:\!NH_3\right]^+$$

with silver ion (or other cation) as an acid and ammonia (or other electron-pair donor) as base. The product of such reaction is often called adduct, a product of a reaction of Lewis acid and base to form a new combination. The following are examples of metal ion complexes: $Cu(NH_3)_4^{2+}$, $FeCl_4^-$, CdI_6^{4-}, the coordination numbers are 4 and 6 respectively. Often coordination centers tend to have a given coordination number which they are associated with.

Generally, coordination centers in solution satisfy their coordination number by forming a complex with the solvent. Therefore, a good example in this case is to take Cu^{2+} ion in aqueous solution which is presented as $Cu(H_2O)_4^{2+}$. In actual fact, complexation reaction is a displacement reaction in which the ligand in the final complex replaces the water ligands associated with the metal ion. This implies that solvent has a significant role to play in complex formation reactions.

Theoretically, any Lewis base can form a complex through coordination with a metal ion; however, the stability of the complex depends on the nature of the reacting substances and on the reaction conditions (temperature, pressure, solvent, etc.). The stronger the base, the greater affinity of the metal ion for an ion pair, the greater will be the stability of the resulting compound.

Generally speaking, most stable complex compounds are formed with transition metals and rare earth elements. In addition, some ligands are capable of complexing the alkaline earth metals (Ca^{2+} and Mg^{2+}). Metal ion follows the rule of adding electrons to fill available electronic orbitals until it reaches or approaches the configuration of the next noble gas following it in the periodic table. For example, Fe^{2+} ion has $26 - 2 = 24$ electrons; the next rare gas, Krypton, has 36; therefore, the iron (II) ion can accommodate 12 additional electrons. It can thus form six coordinate covalent bonds. Also, this applies to Co^{3+} ion. On the other hand, copper (II) ion with 28 electrons, needs (8) eight electrons more to get the Krypton configuration to form four coordinate covalent bonds.

The above is not rigorously followed by all metal ions, since the Co^{2+} ion also coordinates with six donated electrons and by so doing it acquires one more electron than Krypton ($25 + 12$), Ni^{2+} with 26 electrons accommodates either 8 or 12, but not 10. In every case, however, the total number of electrons around the metal ion is close to that of the next noble gas.

Figure 6.1: Complexes with Square Structure

The common and well-known coordination numbers are 4 and 6. Scientific evidence has indicated that coordinating ions or molecules arrange themselves around a metal ion in a defined geometrical configuration. Metals with a coordination number *four* have the coordinating groups distributed either in a *square* (Figure 6.1) or *tetrahedral:* see Figure 6.2 geometry. For example, Ni^{2+}, Al^{3+} and Pt^{2+} have exclusively or predominantly a square planar geometry.

Figure 6.2: Complexes with Tetrahedral Structure

In addition, Be^{2+}, Cu^{+}, Zn^{2+}, Co^{2+}, Cd^{2+} and Mg^{2+} have a *tetrahedral* arrangement. A coordination number of *six* is frequently encountered in complex ions. The geometric configuration of such complex is called *octahedral* (Figure 6.3).

Figure 6.3: Complexes with Octahedral Structure

Most common examples of octahedral complexes in Figure 6.3 are those of the transition metals and rare earth metals. Together with common coordination number 4 and 6, there is monovalent coordination which is rare. Monovalent gold and silver exhibit a coordination number of two. A good example is the complex ions of $Ag(CN)_2^-$, $Ag(NH_3)_2^+$, $AuCl_2^-$, which have a *linear geometry*. It is important to note that coordination numbers 3, 5, 7 and 8 are quite uncommon in complex compounds. The following are good example of each: $KCu(CN)_3$, $Fe(O)_5$ trigonalbipyramid, UF_7^{3-}, and $Zr\ F_7^{3-}$ have a pentagonal bipyramid configuration with a seven coordination number, and TaF_8^{3-} a coordination number of eight.

Metal ions vary greatly in their preferences for the most donor atoms (oxygen and nitrogen). The following in their highest oxidation states coordinate more strongly with oxygen in their highest oxidation states than with nitrogen: Mg, Ca, Sr, Ba, Ga, In, Te, Ti, Cr, Th, Sn, V, Nb, Ta, Mo, U, and Fe. Also, the following seemed to prefer coordination with nitrogen as a donor atom: Cu, Cd, Au, Hg, Co, and Ni. Platinum metal ions, as well as Fe^{2+}, and Cr^{3+}, form strong complexes with both donor atoms (oxygen and nitrogen).

Any ion or molecule which coordinates with metal is called a *ligand*. Ligands are classified based on the number of electron pairs available for coordination or based on chemical nature. Ligands which donate one pair of electrons in the formation of coordinate covalent bonds, which include CN^-, H_2O, NH_3, are called *unidentate* ligands. It should be clear that the number of unidentate ligands per metal ion is equal to the coordination number.

Thus, the copper- ammonia complex has four ammonia molecules per copper ion. Other types of ligands are the *polydentate* or the *chelating* compounds and they can be subdivided as in Table 6.1 hereunder.

Table 6.1: Subdivision of Some Polydentate or Chelating Compounds

Types	Example
(i) Bidentate	
(ii) Tridentate	
(iii) Quadridentates	
(iv) Hexadentate	

The ligands in Table 6.1 can form multiple bonds with a single coordinate center. These type of ligands are referred to as *bidentate* (two bonds), *tridentate* (three bonds), *quadridentates* (four bonds), and *hexadentate* (six bonds). In actual fact, the term *dentate* means that the ligand with multiple bonds is able to "bite" the central ion. The complex formed with a multidentate ligand is called a *chalate*. It should be clearly noted that most complexing agents for suitable complexation titration are *polydentate*.

The stability of a complex, which is a result of covalent bonds formed in the complexiation reactions, is due to the availability of the electron pair of the donor ligand groups. Complexes of oxygen or nitrogen are much stronger than those with other donor elements. Ligands are divided from acidic, basic or amphoteric compounds.

The following are considered as acidic ligands: Cl^-, CN^-, SO_4^{2-} and $RCOO^-$ because they are acidic anions. The resulting charge on the complexes of such anions is the sum of *positive* and *negative* charges of the metal ion and the coordination groups. For example, $Ag(CN)_2^-$, $Fe(CN)_6^{3-}$, and $Cu\,Cl_4^{2-}$. Basic (or neutral) ligands are those in which the coordination occurs directly between an uncharged molecule and the metal ion. The total ionic charge of the complex ion remains uncharged; for example, $[Cu(NH_3)_4]^{2+}$, $[Co(H_2O)]_6^{3+}$ or $[Pt(en)_3]^{4+}$.

The *amphoteric* ligands are polydentates in which some of the coordinating groups are neutral and others acidic; for example, amino acids, (glycine) H_2NCH_2COOH. In neutral solutions, amino acid exists in the form of a "zwitter ion" $^+H_3N\text{-}CH_2COO^-$

The following is a general reaction of amino acid and a Lewis base:

Both the nitrogen and the carboxylic oxygen atom are electron-pair donors and coordinate with a metal ion. Thus,

The resulting complex is called inner complex salts: see Figure 6.4. The latter complex does not conduct electricity in solution because it has no charge. It should always be noted that the positions around metal ion used for coordination are supposed not to be occupied by any ligand. The structure can be seen below.

Figure 6.4: Structure of Inner Complex Salts

Herein is a list of numerous complexes with mixed ligands, a good example of which is ammoniated platinic ion $[Pt(NH_3)_6]^{4+}$, the six ammonia can be replaced by chloride as follows: $[Pt(NH_3)_5Cl]^{3+}$, $[Pt(NH_3)_4Cl_2]^{2+}$, $[Pt(NH_3)_3Cl_3]^+$, $[Pt(NH_3)_2Cl_4]^0$, $[Pt(NH_3)Cl_5]^-$, and $[PtCl_6]^{2-}$.

6.3   Stability of Complex Compounds

Strictly speaking, we cannot consider any metal ion in solution as uncomplexed, since its coordination positions will be occupied by the solvent molecules. The ligand does not add to the metal ion, but replaces the solvent. In dilute solutions, where it is always considered that the activity of the solvent as unity, this does not affect the equilibrium constants given below. When a ligand A is added to a solution of a metal salt capable of forming a complex, equilibrium is established between the uncomplexed metal and the ligand as can be seen below:

where ion M reacts with p ions or molecules of the ligand.

The *formation* or stability constant of the complex can be calculated by the expression 6.1.

$$K_f = \frac{MA_p}{(M)\,(A)^p} \tag{6.1}$$

whereas the quantities in parentheses are the equilibrium activities of the reacting species. In addition, the *dissociation* or instability constant of the complex is obtained by expression 6.2.

$$K_d = \frac{(M\,A)^p}{(MAp)} = \frac{1}{K_f} \tag{6.2}$$

Thus, the smaller the dissociation constant, the greater is the stability of the complex. For example, $[Ag(NH_3)_2]^+$, $K_d = 6.8 \times 10^{-8}$ is less stable than $[Ag(CN)_2]^-$, which has a dissociation constant of $K_d = 1.0 \times 10^{-21}$.

It should be clear that the formation constant K_f in expression 6.1 is the overall formation constant for the complexation reaction. Since the reaction takes place in a stepwise manner, the individual constants are obtained in the following sequences:

$$K_1 = \frac{(MA)}{(M)(A)} \tag{6.3}$$

$$K_2 = \frac{(MA_2)}{(MA)(A)} \tag{6.4}$$

$$K_p = \frac{(MA_{p-1})}{(MA_{p-1})\,(A)} \tag{6.5}$$

and, $K_f = K_1.K_2.K_3 \ldots \ldots K_p$: $\tag{6.6}$

For various reasons this discussion cannot go into that much detail about the factors which contribute to the stability of a complex compound; instead it will only focus on general statements concerning stability or instability, as follows:

(i)   The most stable complexes are formed with the small cations of the transition metal elements. The order of increasing stability for divalent transition ions is as follows: Mn < Fe < Co < Ni < Cu > Zn.

The large cations of the alkali and alkali earth metals form complexes that are less stable than those formed by transition metal ions.

(ii)   The stability of complexes containing halogen or pseudo-halogen ligands decreases in the following order $CN^- > SCN^- > F^- > Cl^- > Br^- > I^-$.

The order of decreasing stability for unidentate ligands containing the same or different donor atoms is as follows: $NH_3 > RNH_2 > R_3N > H_2O > ROH > R_2O > R_3As > R_3P > R_2S$, and

(iii)   Ring formation in coordination reactions leads to increased stability in the resulting complex.

Example 6.1

A 0.1 M solution of $KAg(CN)_2$ is prepared in water. Assuming that the salt is completely ionized to K^+ and $Ag(CN)_2^-$ ions and that the dissociation constant of the latter complex ion is 4×10^{-21}, calculate the concentration of the free silver ion in solution. This does not consider activity coefficients.

Solution:

The dissociation equilibrium is

$$Ag(CN)_2^- \;\rightleftharpoons\; Ag^+(aq) \;+\; 2\,CN^-(aq)$$

$$K_d = \frac{(Ag^+)(CN^-)^2}{(Ag(CN)_2)} = 4 \times 10^{-21}$$

Let x be the concentration of free silver ion in the solution, then the concentration of the free cyanide ion will be $2x$ and that of the complex is $0.1 - x$. Therefore,

$$\frac{(x)(2x)^2}{0.1 - x} = 4 \times 10^{-21}$$

This being a cubic equation, x will be solved by successive approximations, assuming that $0.1 \gg x$

$4\,x^3 = 4 \times 10^{-21}$

$x = 0.7 \times 10^{-7} \sim 10^{-7}$

Example 6.2

A complex compound MA_3 is dissolved in water to form a 0.20 M solution. A colorimetric measurement indicates that the concentration of free "A" is 0.30 M. Assuming that the dissociation of the complex occurs in one step, calculate its formation constant. Ignore activity corrections.

Solution:

The following is the dissociation reaction

and the formation constant is given by

$$K_f = \frac{(M\,A_3)}{(M)(A)^3}$$

If the concentration of free "A" is 0.03 M, the corresponding concentration of free "M" must be 0.01 M, while that of complex is $0.20 - 0.01 = 0.19$

$$K_f = \frac{(0.19)}{0.01(0.03)^3} = \frac{0.19}{1.7\ x\ 10^{-7}} = 7.0\ x\ 10^5$$

Also from complexation reaction expression 6.7, the effective formation constant $K_{eff\ in}$ expression 6.8 is always important, particularly when you are dealing with titration curves of complex materials, as follows:

$$A \quad + \quad B \; \rightleftharpoons \; AB \tag{6.7}$$

$$K_{AB\ eff} = \frac{(AB)}{(A)(B)}) \tag{6.8}$$

6.4 Analytical Applications of Complex Compounds
6.4.1 Dissolution

The formation of complex ions is one of the essential steps in the process of dissolution. Many metals, metal oxides, and salts can be dissolved in aqueous solutions through the formation of soluble complex ions. Such formation is due to the coordination of inorganic anions with metallic cations. For example, the halides and the hydroxyl ion are the ligands which form large class of such soluble metal complexes. The fluoro complexes are commonly used in dissolution process. A good example is Niobium, Tantalum, Zirconium, and the oxides of these metals dissolve in hydrofluoric acid (HF) due to the formation of fluoro complex.

Chloro complexes are also useful in solubilizing various materials. For example, the solubility of AgCl occurs in solution containing excess chloride ion. Likewise, lead sulfate can be dissolved in HCl with the formation of the soluble $PbCl_4^{2-}$ ion. Indeed, gold and platinum alloys are dissolved in *aqua regia* because in part there is formation of chloro complexes. Ammonia dissolves silver halides with formation of the soluble silver ammine complex, $Ag(NH_3)_2^+$. In fact some strong bases dissolve certain metals by forming soluble complex ions. For example, Al and Zn can be dissolved in NaOH with formation of aluminate and zincate ions, respectively (expression 6.9 and 6.10).

$$2\,Al \;+\; 2\,NaOH \;+\; 2\,H_2O \longrightarrow 2\,NaAlO_2 \;+\; 3\,H_2 \tag{6.9}$$

$$Zn \;+\; 2\,NaOH \longrightarrow Na_2ZnO_2 \;+\; H_2 \tag{6.10}$$

6.4.2 Separations

Complex ions have a significant role to play in analytical separations. Some of these separation techniques we will see them later. The coordination of a metal ion with an amphoteric organic ligand, which reacts in its anionic form, usually results in a decreased water solubility of the metal ion. For example, dimethylglyoxime forms complexes with nickel, Ni; palladium, Pd; platinum, Pt; copper and divalent silver. Only Ni and Pd complexes are insoluble in water. Since Pd is not precipitated in dilute ammonical solutions, dimethylglyoxime is a very selective precipitating agent for nickel in basic solution, as can be seen in expression 6.11.

$$Ni^{2+} + 2 \begin{array}{c} H_3C-C=NOH \\ | \\ H_3C-C=NOH \end{array} \longrightarrow 2 H^+ + \quad (6.11)$$

Furthermore, Pd is quantitatively precipitated under acidic conditions. Some organic precipitating agents show high selectivity; for example, 8-hydroxyquinoline is commonly used as a precipitant for aluminum as can be seen in expression 6.12.

$$Al^{3+} + \quad \longrightarrow \quad + \quad H^+ \quad (6.12)$$

Not only has o-hydroxybenzaldehyde some selectivity for copper expression 6.13, but it will also precipitate both Pd and vanadium.

$$1/2\ Cu^{2+} + \quad \longrightarrow \quad + \quad H^+ \quad (6.13)$$

Though 1-nitros-2-naphthol is fairly specific for cobalt, it forms insoluble precipitates with Cu, Fe, Pd, Vanadium and zirconum. The isolated colbalt complex has the metal in +3 oxidation state. Indeed, it has been proposed that both complexation and oxidation occur in two step reaction as can be seen in expressions 6.14 and 6.15 below:

$$\tfrac{1}{2}\,Co^{2+} \; + \quad\longrightarrow\quad \; + \; H^+ \qquad (6.14)$$

$$2\,Co\,X_2 \;\longrightarrow\; CoX_3 \;+\; Co^{2+} \;+\; X' \quad \text{(X' is the reduced form of the reagent)} \qquad (6.15)$$

It should be noted that cupferron (the ammonium salt of nitrosophenylhydroxylamine) and neocupferron (the ammonium salt of nitrosonaphylhydroxylamine) are good precipitants for iron Fe, niobium Nb, tantalum Ta, tungsten W, and zirconium Zr, in very strong acidic solutions. The previous elements are also precipitated in dilute hydrochloric acid or sulfuric acid, including gallium Ga, tin Sn, antimony Sb, palladium Pd, as well as titanium Ti. The rare earth metals are often precipitated between 3-4 pH ranges, but not in a more acidic environment; however, aluminum is precipitated out in a neutral solution.

From the fact that neutral salts are formed with the above mentioned reagents, this facilitates the extraction of salts from water into a nonaqueous medium to become easily achieved. Common immiscible solvents such as benzene, carbon tetrachloride, chloroform, hexanol and nitrobenzene are the ones used to remove these complexes from aqueous solutions. For example, nickel dimethylglyoximate complex is extracted from aqueous solution with chloroform. Also metal oxine complexes (8-hydoxyquinolates) are extracted with chloroform as the cupferrates and neopcupferrates. In addition, chloroform extracts metal complexes of diphenylthiocabazone.

Some organic ligands which react as neutral molecules in coordination reactions with metal ions do not most of the time cause precipitation of the metal ion. There are some incidences where subsequent association of the metal cation complex with an anion form an ion associated complex leading to insolubility and extractability. A good example is the soluble, blue copper-pyridine complexes which often form an insoluble thiocyanate, $CuPy_2(SCN)_2$, which can be extracted with chloroform. Furthermore, the ferrous phenanthroline complex can be precipitated in the form of iodide or perchlorate salt, as can be seen in the structure below.

The previously mentioned salts can be extracted from water by using isoamyl alcohol or nitrobenzene. It should be noted that both red salt bis(2,2-bispyridine) copper(I) perchlorate and tris(2,2′-bipyridine)ion(II) perchlorate are water insoluble; however, they are extremely soluble in isoarmyl alcohol or nitrobenzene.

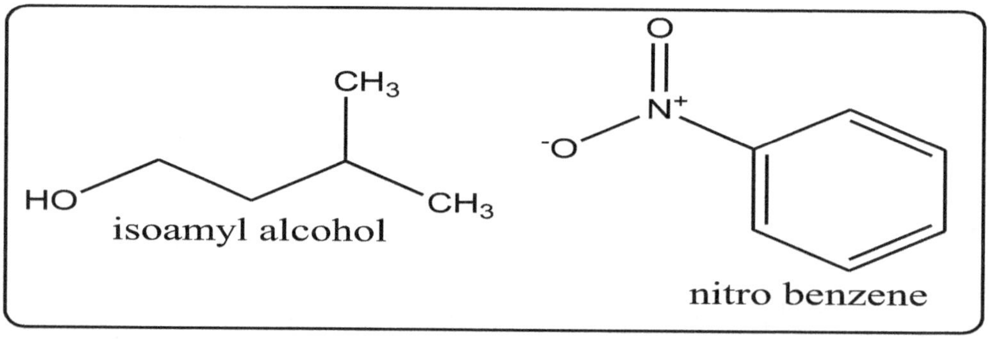

Another useful example is the complex with the following general formula MY^-_{n+1}, where Y is a univalent anion and n is the oxidation number of the metal ion. For such a case, complexation is achieved through an anionic association with the metal ion. Anionic complex association with cationic often results in the formation of neutral salts which can be extracted from water. A common example is the one resulting from the reaction of iron with concentrated hydrochloric acid in which the $FeCl_4^-$ is formed. The association of $FeCl_4^-$ with hydrogen ion results in the formation of a neutral specie, $HFeCl_4$, which can be extracted from aqueous solutions with ether. Obviously, there are many other bromo, chloro, iodo, and thiocyanato complexes which are also useful in extracting metal ions out of aqueous solution, though they are not discussed in this section.

Complex compounds are also very important in ion exchange separation. Conversion of cationic species to anionic species by complexation is a very common way of increasing the separability of metal ions. In a mixture of calcium and iron, for example, reaction with hydrochloric acid would give the $FeCl_4^-$ ion which would be retained by the anionic ion exchange resin. The reason for this phenomenon can be obtained by revising chapter four. Moreover, calcium ion would not be held by an ion exchange column, and thus the separation would be achieved. There are several numbers of coordinating ligands that are used in effecting separation by ion exchange. The simple halogen and pseudo-halogen ions have proven to be more effective than sophisticated systems.

To a certain extent, electrolytic separations can be achieved through complex formation. As is always the case, complexation of a metal ion results in a shift of the reduction potential to more negative values and this affects the rate of electron exchange between electrode and the electroactive species. There are always two possibilities: a complex species may exchange electrons faster or slower than the simple metal ion. The chances are that the two effects may be in the same direction or may oppose one another. The literature shows that the simple *aquo* complexes of most metal ions are reduced more easily than corresponding metal complex ions. The reduction potential of Ag^+ and $Ag(CN)_2^-$ ions are such that the cyano complex ($Ag(CN)_2^-$) is relatively more difficult to reduce to metallic silver than the simple ion (Ag^+). The deposit of silver from silver deposition is very smooth and adheres tightly to the cathode. This is a special advantage in the electrolytic separation of silver.

6.4.3 Masking

This is a process whereby reactive interfering metal ions are held by complex formation reagents and become practically unreactive with titrant. Such a characteristic is important because interfering substances may react with added reagent or with the reaction product, and thus render the determination difficult or impossible. The masking reagent will change the chemical nature of interfering metal ions; though it will remain in the reaction mixture, it will no longer interfere with main analytical reaction. Good examples of masking agents include citrate, cyanide, fluoride, triethanolamine, 2,3-Dimercaptopropanol and phosphate ions. These ions form stable complexes with iron, and other metal ions. The fluoride ion is commonly used to mask iron in the titrimetric of copper in copper ore as well as Al^{3+}, Fe^{3+}, Ti^{4+}, and Be^{2+} ions. Furthermore, cyanide is also a masking agent that forms complexes with Cd^{2+}, Zn^{2+}, Hg^{2+}, Co^{2+}, Cu^+, Ag^+, Ni^{2+}, Pd^{2+}, Pt^{2+}, Fe^{2+}, but not with Ca^{2+}, Mg^{2+}, Mn^{2+}, or Pb^{2+}. This means when CN^- ions are added to a solution containing Cd^{2+} and Pb^{2+} only Pb^{2+} can react with EDTA. Point of caution: Cyanide forms toxic gaseous HCN below pH 11, and because of the latter drawback cyanide solution should be made strong basic and handled in a fume hood. Triethanolamine masks Al^{3+}, Fe^{3+}, and Mn^{2+} while 2,3-Dimercaptopropanol masks Bi^{3+}, Cd^{2+}, Cu^{2+}, Hg^{2+}, and Pb^{2+}. There are several applications of masking reactions in analytical determinations that can be cited. For example, a technique used to separate nickel from steel involves the precipitation of nickel by dimethylglyoxime as an insoluble nickeldimethylglyoxime complex. Nevertheless, the precipitation should be carried out in a slightly alkaline solution, at a pH where all the iron would precipitate as hydrated oxide. A difficult situation is avoided by the addition of tartaric acid to the solution. The tartrate ion formed leads to the formation of soluble complex with iron, and thereby solubilizes ferric hydroxide. This means the addition of tartrate ion severely decreases the concentration of the free Fe(III) ions to the point where it makes a concentrated solution of the base to exceed the solubility product of Fe(III) hydroxide.

An obvious and well known common masking agent is EDTA. The latter being hexadentate ligand, coordination occurs in a single step even with metal ion having coordination number of six. Alkaline earth metals, transition metals, and rare earth metals form complex with the ligand in its anionic forms of H_2Y^{2-} or HY^{3-}. Complexation of heavy metal ions with EDTA prevents hydroxide precipitation, thus preventing possible sources of interference if the analysis is carried out in basic solution.

In addition, masking agent can be employed in titration of Cd^{2+} in the presence of Zn^{2+}. In the case when ammonia and ammonium ion buffer is used, an ammonia-zinc complex will be formed. The latter effect greately reduces the effective formation constant for the zinc ion. This approach is a good example of a secondary differentiating characteristic which circumvents interference. In fact selectivity afforded by masking and pH control allows each individual component of complex mixtures to be analyzed by EDTA titration.

HO   SH

SH

2,3-Dimercaptopropanol

OH   O

HO   OH

O   OH

tartaric acid

Example 6.3

(a) A 50.0 mL sample containing Ni^{2+} was treated with 25.00 mL of 0.050 M EDTA to complex all the Ni^{2+} and leave excess EDTA in solution. How many millimoles of EDTA are contained in 25.0 mL of 0.050 M EDTA?

(b) The excess EDTA in part (a) was back-titrated, requiring 5.00 mL of 0.050 M Zn^{2+}. How many mmol of Zn^{2+} are in 5.00 mL of 0.050 M Zn^{2+}?

(c) The millimoles of Ni^{2+} in the unknown is the difference between the EDTA added in part (a) and the Zn^{2+} required in part (b) Find the number of millimoles of Ni^{2+} and the concentration of the Ni^{2+} in the unknown.

Solution:

(a) (25.00 mL) (0.050 M) = 1.25 mmol EDTA

(b) (5.00 mL) (0.050 M) = 0.25 mmol Zn^{2+}

(c) mmol Ni^{2+} = mmol EDTA - mmol Zn^{2+}

$$1.25 - 0.25 = 1.00 \text{ mmol } Ni^{2+}$$

$[Ni^{2+}]$ = (1.00)/(50.00 mL) = 0.02 M Ni^{2+}

6.4.4   Measurements

(i) *Gravimetry*: In gravimetric methods of analysis, complexes formed between metal ions and organic ligands are extremely useful. Organic precipitating agents have more advantages than inorganic reagents. Organo-metallic precipitates are often bulky, easy to filter if they are stable, easy to dry, and give advantageous weight ratios. In most cases organic precipitating agents are not highly selective, but their selectivity is enhanced through controlling of the individual solution pH.

The organometallic precipitate can be filtered, washed, and either dried at 100-110 °C or ignited to the metallic oxide. For example, the nickel-dimethylglyoxime complex is stable at 110 °C, and is not appreciably contaminated by co-precipitation. Thus, the complex itself serves as easy form of weighable nickel.

(ii) *Titrimetry*: Titration of a metal ion with a suitable electron donor ion or molecule can be regarded as a Lewis acid – base titration. Generally speaking, there are no significant differences between expressions 6.17 and 6.18 shown below:

$$K_f = \frac{(H_2O)}{(H^+)(OH^-)} \tag{6.16}$$

$$K_f = \frac{(HCN^-)}{(H^+)(CN^-)} \tag{6.17}$$

$$K_f = \frac{(CuCN^+)}{(Cu^{2+})(CN^-)} \tag{6.18}$$

The above complexation reactions can be adapted to analytical reaction when they proceed *rapidly* and basically *completely* in dilute solution. Such reactions must be stoichiometric; that is, the reaction products must be with clear defined chemical species which have known and constant composition. Lastly, there should be a suitable indication of the equivalence point during titration. The issue of fast and essentially complete reactions is of paramount importance. Luckily enough, most complexation reactions often proceed sufficiently rapidly to meet the requirement of speed. Trivalent metal ions, for example chromium, cobalt, and iron have some exceptions in coordination reactions. Attempting to replace coordinated water molecules in these ions by using organic ligands is relatively slow.

Thus, the latter reactions are not recommended for titrimetric analysis. As we have previously seen in Chapter 2, basically all reactions in solutions are to some extent, at least, reversible and do not go to completion. For analytical purposes, a reaction is considered suitable if, at equilibrium, the extent of the reaction is not less than 99.9%.

Example 6.4
Assume that in a complexometric titration we are reacting 0.1 M solution of a metal ion M with a 0.1 M solution of a ligand A.

Solution:
The reaction equation is

and the formation constant

$$K_f = \frac{(MA)}{(M)(A)} \tag{6.19}$$

From the fact that equal volumes of two solutions must be mixed in order to arrive at the proper stoichiometric proportion, the concentration of MA in solution would obviously be 0.05 M. With the assumption that the reaction must be at least 99.9%, the *minimum* concentration of MA should be 4.9995 x 10^{-2} M and the *maximum* concentration of free M and A should be 5 x 10^{-5} M

each. Therefore, if substitution is done in expression 6.19 the *minimum* value for the formation constant K_f *must* be as follows:

$$K_f = \frac{(4.995 \; x \; 10^{-2})}{(5 \; x \; 10^{-5})(5 \; x \; 10^{-5})} = 2 \; x \; 10^7$$

In general, a smaller K_f means that the reaction is *less* than 99.9% complete at the equivalence point, and therefore does not qualify for analytical work. Usually, high stability constant values are needed if the titration is to be carried out in more dilute solution.

Complexometric titrations of unidentate ligands are not generally useful, although in some cases complete reactions may be achieved. This turned out to be true due to the fact that the replacement of coordinated water molecules in the coordinated sphere of a metal ion by the titrant ligand or molecule occurs in a stepwise style: see expressions 6.3 up to 6.5. During titration there are series of different complex species which are formed in solution. Most of these may exist simultaneously in the titration mixture, due to successive complexation overlap. Stoichiometric relationships are obscure and provide an ill-defined titration curve. Such a situation looks similar to the titration of a weak polybasic acid in which inflections in the titration curve are always difficult to observe: see Figure 6.5.

Polydantate ligands forming very stable complexes are well known to analytical chemists. Some of these ligands undergo coordination in single step. Common compounds with such characteristics include iminodiacetic acid, $NH(CH_2COOH)_2$, Nitrilotriacetic acid, $N(CH_2COOH)_3$, and especially EDTA are compounds which form very stable complexes in single step with various metal ions. A good example is the formation constants of EDTA complexes with Ca^{2+}, Cu^{2+} and Zn^{2+} which are, respectively, $4 \; x \; 10^{10}$, $3.2 \; x \; 10^{18}$, and $4 \; x \; 10^{16}$.

It is evident that titration involving EDTA experiences competition between metal ion and hydrogen for organic ligand. Below is the EDTA, H_4Y dissociation representation.

$$H_4Y \rightleftharpoons H^+ \; + \; H_3Y^- \; ; \qquad\qquad K_1 = 1.0 \; x \; 10^{-2}$$

$$H_3Y^- \rightleftharpoons H^+ \; + \; H_2Y^{2-} \; ; \qquad\qquad K_2 = 2.0 \; x \; 10^{-3}$$

$$H_2Y^{2-} \rightleftharpoons H^+ \; + \; HY^{3-} \; ; \qquad\qquad K_3 = 6.3 \; x \; 10^{-7}$$

$$HY^{3-} \rightleftharpoons H^+ \; + \; Y^{4-} \; ; \qquad\qquad K_4 = 5.5 \; x \; 10^{-11}$$

The approximate distribution of EDTA species as a function of pH is shown in Figure 6.1. In Figure 6.5 the reagent exists almost completely at around pH 1 or less and as Y^{4-} at pH 12 or higher. Between pH 4 and 5 the reagent exists as H_2Y^{2-} and it is almost entirely as HY^{3-} in the pH range 8-9. A pH 10 is common pH titration of calcium ion with EDTA; however, almost 67% of the reagent exists as HY^{3-} and 33% exists as Y^{4-}.

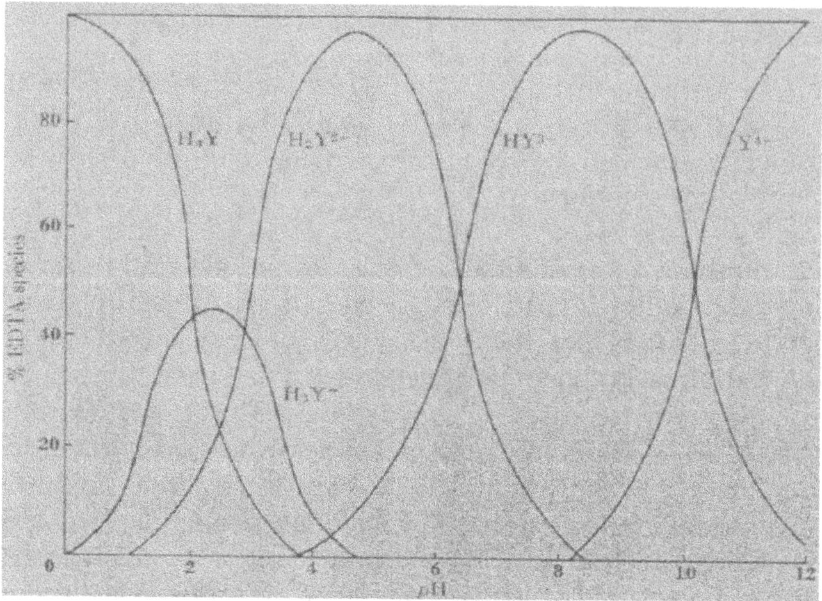

Figure 6.5: Distribution of EDTA as Function of pH

Let us consider titration of calcium ion with a standard solution of EDTA. The complexation reaction in a system buffered at pH 10 (see Figure 6.6) is nearly quantitative and occurs in a single step, as follows:

In titration a metallic indicator is used to detect the end point. During titration, the indicator competes with the titrant for the calcium ion. If Eriochrome black T, represented below, is used as the metalchromic indicator, the presence of a small amount of magnesium improves the colour change in the titration.

Below is a typical well-defined titration curve for calcium.

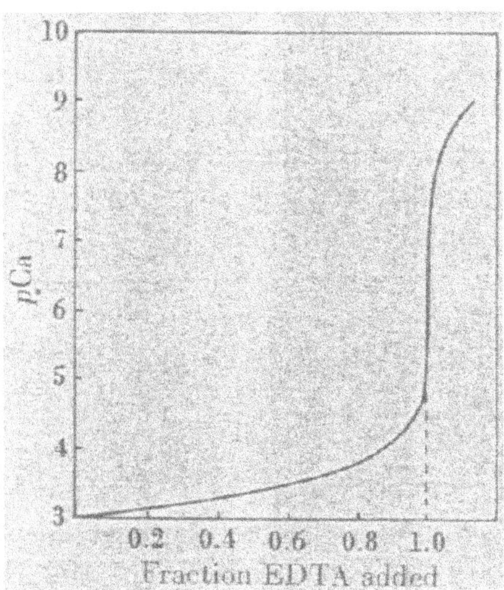

Figure 6.6: Complexometric Titration Curve with EDTA as Titrant

This is a very important discussion on comploximetric titration and cannot end without looking at how cyanide ion is titrated by a standard solution of silver nitrate. Upon addition of excess of cyanide ion to a silver ion, the soluble complex $Ag(CN)_2^-$ is formed. Once all cyanide has been complexed by silver ion, just a slight addition of the latter produces an insoluble or slightly soluble silver cyanide salt expression 6.20.

$$Ag^+ \quad + \quad Ag(CN)_2^- \quad \rightleftharpoons \quad Ag^+(Ag(CN)_2^- \qquad (6.20)$$

Generally, the above precipitate serves as an end point in the titration. The sharp end point in the latter can be obtained by a small addition of iodine ion to the cyanide solution. Once the entire dicyanosilver ion has been formed, a small excess of silver ion produces a precipitate of silver iodide which is more insoluble and relatively more visible than that of silver cyanide.

The titration of nickel with a standard solution of potassium cyanide is another application of complexation of a metal with cyanide ion. Just an addition of a small amount of silver iodide as an indicator in the reaction of ammonical solution is as shown in expressions 6.21 and 6.22.

$$Ni(NH_3)_4^{2+} \quad + \quad 4\,CN^- \quad \rightleftharpoons \quad Ni(CN)_4^{2-} \quad + \quad 4\,NH_3 \qquad (6.21)$$

$$AgI \quad + \quad 2\,CN^- \quad \rightleftharpoons \quad Ag(CN)_2^- \quad + \quad I^- \qquad (6.22)$$

The disappearance of turbidity caused by the insoluble silver iodide signifies the end point.

6.5 Quantitation by Complexation Titration

Complexation reaction is often used for quantitative analysis by titration in an approach similar to the one which we will see later in chapter eight under acid-base reactions. For such types of reactions a suitable complexing titrant is chosen for the species to be titrated. Most of the time, the analyte is a coordination center, and the titrant is the ligand of complexing agent. There are many techniques for determining the equivalence point on the curve of such a reaction. For example, an electrode whose potential is dependent on the coordination center (or ligand) concentration can be used. Also, in the case where reactant of the complex formed possesses a significant absorbance, then absorbance spectrometry is recommended.

6.5.1　Achieving Selectivity with EDTA Titration

One important advantage of EDTA, among others, is a broad range of metal ions which it can complex. This characteristic makes it considered as a good-purpose reagent. On the other hand, the ability of a metal to react with EDTA is not a highly differentiating characteristic. Obviously, many metal ions other than the desired analyte are potential interferents. To avoid this problem, advantage can be taken to the great range of effective formation constants (expression 6.8) for different metal ions. These are well shown in Figure 6.7. It should be noted that effective formation constant is significantly affected by $p[H_3O^+]$ because of the protonation of EDTA and the hydrolysis of the metal ion.

A number of techniques can be used, for example, to titrate Hg^{2+} in the presence of Cd^{2+}, and by just setting $p[H_3O^+]$ at 2.00 it will remove the chances of the formation of cadminium complex, while allowing the mercury complex formation constant to remain high. In a situation where interfering ion cannot be easily removed by the choice of $p[H_3O^+]$, a $p[H_3O^+]$ can always be found where K_{eff} has enough differences for the titration curve to form clear separate breaks for each metal. It has been found that at least a difference of four or more in $\log K_{eff}$ is enough for this type of separation to happen.

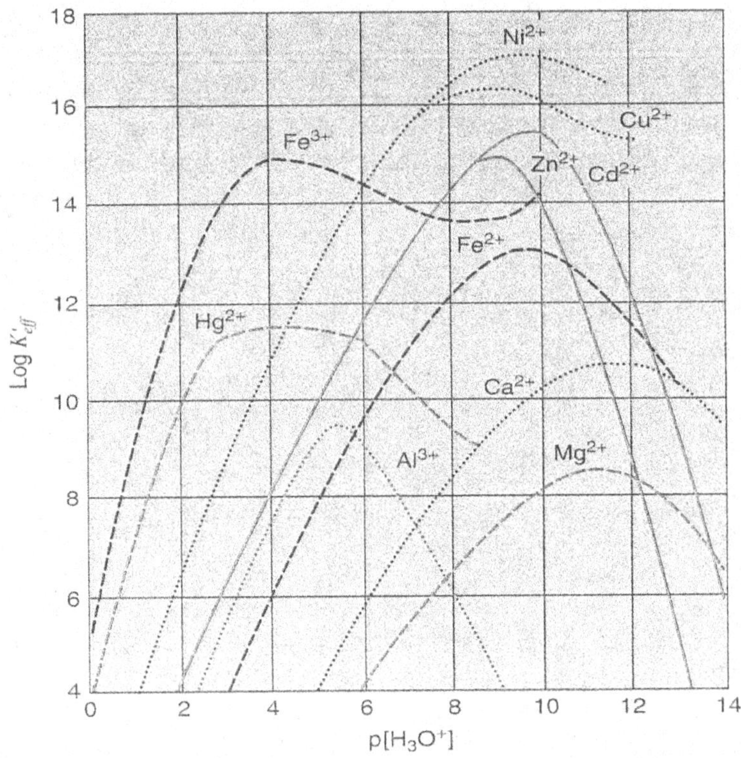

Figure 6. 7: Log K'_{eff}' versus p[H$_3$O$^+$]

6.6 Problems

6.6.1. Outline three desirable properties of a gravimetric precipitate.

6.6.2. Why is reprecipitation employed in a gravimetric analysis?

6.6.3. With a diagram draw the general structure for coordination number four geometry.

6.6.4. Explain why inner complex salts do not conduct electricity.

6.6.5. Provide descriptions on how you are going to separate Ni ions from the mixture of Pd and Pt ions in solution.

6.6.6. What are the qualities that an indicator for complexation titration should have?

6.6.7. Why is the presence of Mg^{2+} desirable when titrating Ca^{2+} with EDTA using Eriochrome Black T as an indicator?

6.6.8. How can Mg^{2+} be added to the solution without affecting the volume of EDTA needed to titrate the Ca^{2+}?

6.6.9. Why are chelates rather than single coordination site ligands chosen for complexometric titration?

6.6.10. How does the magnitude of the concentration change in the equivalence point region affected by an increase in the formation constant of the complex and by a decrease in the concentration of analyte?

6.6.11. (a) In order to measure nickel content in steel, the alloy should be dissolved in 12 M HCl and neutralized in the presence of citrate ion, which often maintains iron in solution. The slightly basic solution is warmed, and dimethyglyoxime (DMG) is added to precipitate the red DMG-nickel complex quantitatively. The product is filtered, washed with cold water, and dried at 110 °C. If the nickel content is known to be near 3 wt% and you wish to analyze 1.0 g of the steel, what volume of 1.0 wt% alcoholic DMG solution should be used to give a 50% excess of DMG for analysis? Assume that the density of the alcohol solution is 0.79 g/mL.

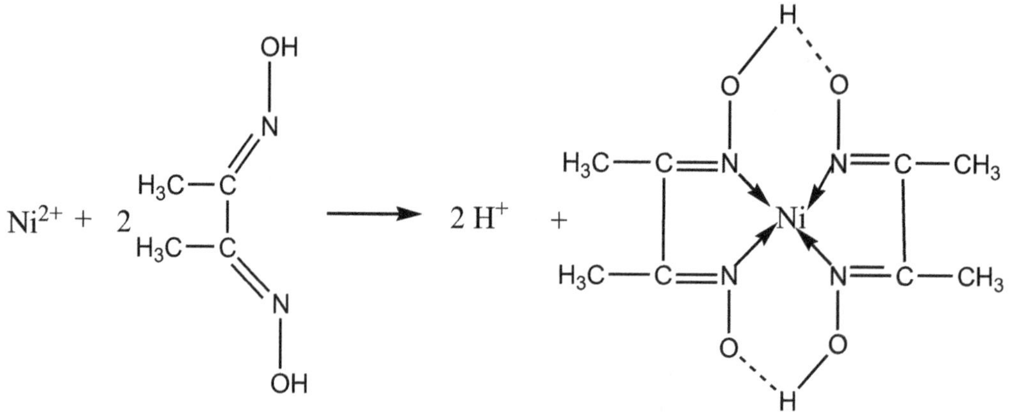

$$Ni^{2+} + 2\;DMG \longrightarrow 2\;H^+ + Ni(DMG)_2$$

(b) If 1.163 g of steel gave 0.179 g of precipitate, what is percentage of Ni in the steel?

6.6.12. An organic compound with a formula mass of 417 g/mol was analyzed for ethoxy (CH_3CH_2O-) groups by the following reactions below.

$$ROCH_2CH_3 + HI \longrightarrow ROH + CH_3CH_2I$$

$$CH_3CH_2I + Ag^+ + OH^- \longrightarrow AgI + CH_3CH_2OH$$

A 25.42 mg sample of compound produced 29.03 mg of AgI. How many ethoxyl groups are there in each molecule?

6.6.13. 0.649 g sample containing only K_2SO_4 (FM 174.27 g/mol) and $(NH_4)_2SO_4$ (FM 132.14 g/mol) was dissolved in water and treated with $Ba(NO_3)_2$ to precipitate all the SO_4^{2-} as $BaSO_4$ (FM 233.39 g/mol). Find the weight percent of K_2SO_4 in the sample if 0.977 g of precipitate was formed.

6.6.14. How could one determine the concentration of Fe^{3+} in the presence of Fe^{2+} with an EDTA titration? Explain.

6.6.15. What is the characteristic of the metal ion that determines the position of the left side of its log K'_{eff} curve in Figure 6.7?

6.7 References

Chritiany, G. D. (1994). *Analytical Chemistry*, 5th ed. John Wiley and Sons, NewYork.

Claudia, P. M., Cristhiane, N. A., Dhizuo, M., Marcela, G., and Luziane, F. (2011). Complexometric Titration with Potenciometric Indicator to Determination of Calcium and Magnesium in Soil Extracts, *R. Bras. Ci. Solo. 35*, 1331

Kolthoff, I. M. (1994). "Analytical Chemistry in the U.S.A. in the First Quarter of this Century," *Anal. Chem. 66,* 241A.

Litimer Jr., G. W, (1966). "Piperizine as the Diacetate." *J. Chem. Ed. 43*, 148; G. R. Bond, *Anal. Chem. 32*, 1332.

Martell, A. E. and Honcock, R. D. (1996). *Metal Complexes in Aqueous Solution*, Plenum Press, New York.

Mossman, D. M. Kooser, R. G. and Welch, L. E. (1996). "The Complexometric Determination of Calcium and Magnesium in Limestone Using a Laser Photometer for Endpoint Identification", *J. Chem. Ed. 73*, 82.

Novick, S. F. (1997). "Complexometric Titration of Zinc," *J. Chem. Ed., 74*, 1463.

Richards, T. W. (1925). "Classical Gravimetric and Analysis," *Chem. Rev, 1*, 1.

Teslyuk, O. L., Beltyukova, S. U. and Yegorova, A. V. (2007). Complex Compounds of Terbium(III) with Some Nonsteroidal Anti-inframmatory Drugs and their Analytical Applications, *J. Anal. 62*, 330.

Yatsimirskii, K. B. (2012). *Instability Constants of Complex Compounds*, Pergamon Press LTD.

Chapter 7

SAMPLING

7.1   Introduction

Sampling is the process of collecting representative sample for analysis. The actual samples require to some degree sample preparation to remove substances that interfere in the analysis of the desired analyte. This process will undoubtedly convert the analyte into a form suitable for analysis. In Figure 7.1 a list of important terminologies one may encounter in sampling and preparation are well shown hereunder.

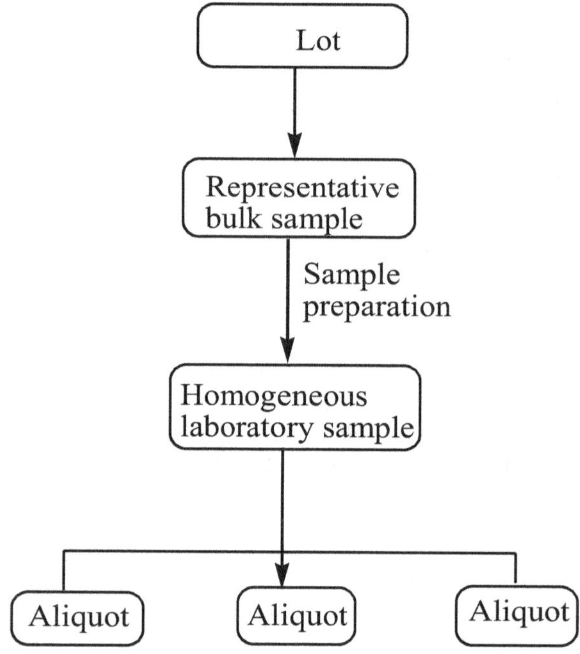

Figure 7.1: Terminology of Sampling

7.2   What is a Lot?

A lot is the total material from which samples are taken. Bulky sample (also called a gross sample) is taken from the lot for analysis or archiving (storing for future reference). The bulk sample must be representative of the lot, and the choice of bulk is critical to producing a valid analysis that gives a strategy for sampling of heterogeneous materials.

From a smaller representative of the bulk, a homogeneous sample is formed that must have the same composition as the bulk sample. For example, one can obtain a laboratory sample by grinding an entire solid bulk sample to a fine powder, mixing thoroughly, and keeping one bottle of powder for testing. Small portions (called aliquots) of the laboratory sample are used for individual analysis. Sample preparation is the series of steps needed to convert a representative bulk sample into a form suitable for chemical analysis.

7.2.1   Where Should we Store Samples?

The sample composition changes with time after collection due to chemical changes, as a result of reaction with air or interaction of the sample with its container. The glass container is a notorious ion exchanger see details in chapter 4, section 4.4.9 that can alter the concentrations of trace ions in solution. Therefore, plastic, especially Teflon, collection bottles are commonly recommended. Even with the use of the latter material they can still absorb trace levels of analytes. For example, a 0.2 µM $HgCl_2$ solution lost 40-95% of its concentration in 4 hours in polyethylene bottles. A 2 µM Ag^+ solution in a Teflon bottle lost 2% of its concentration in a day and 28% in a month.

Plastic containers must be washed thoroughly well before use. The Table 7-1 below shows how manganese in blood serum samples increased by a factor of seven (7) when stored in unwashed polyethylene containers prior to analysis.

Table 7.1: Amount of Manganese in Blood Serum Increased When Stored in Unwashed Polyethylene Container

Container	Mn (ng/mL)
Unwashed	0.85
Unwashed	0.55
Unwashed	0.20
Unwashed	0.67
Average	0.57 ± 0.27
Washed	0.096
Washed	0.018
Washed	0.12
Washed	0.10
Average	0.084 ± 0.045

A washed container was rinsed with distilled water twice from quartz, which introduced less contamination into water compared to a glass container. Steel needles are an unavoidable source of metal contaminants or contamination in biochemistry/chemistry analysis.

Experience has shown that trace analysis is always accompanied with a number of challenges. A study of handling techniques for the analysis of lead in a river revealed variations in sample collection, sample containers, protection during transportation from the field to the laboratory, filtration techniques, chemical preservatives, and preconcentration procedures.

Each step which deviated from known practice doubled the apparent concentration of lead in the stream of water. Clean rooms with filtered air suppliers are essential in trace analysis. Unless the complete history of any sample is known with certainty; the analysist is well advised not to spend his or her time in analyzing it. In the laboratory note book a student should describe how a sample was collected, stored, and exactly handled, as well as stating how it was analyzed.

7.3   Sampling

Suppose you intend to make a universal statement about the amount of "caffeine in chocolate": this will require you to analyze several samples of each type to determine the range of caffeine content in each kind of chocolate from the same manufacturer. It is fine to assume that a pure chocolate bar is *homogeneous*, which means that its composition from one end is the same at the other end. Chocolate with nut in the middle is a good example of *heterogeneous* material, which means that the composition differs from one place to another, because the nut is different from chocolate. This means if you were sampling a heterogeneous material, you will be required to use a different strategy than that used to homogeneous sample material.

7.4   Statistics of Sampling

For random errors, the overall variance, S_o^2 is the sum of the variance of the analytical procedure, S_a^2, and the variance of the sampling operation, S_s^2.

Additivity of variance $\rightarrow S_o^2 = S_a^2 + S_s^2$ (7.1)

When S_a or S_s is sufficiently smaller than the other, there is little point in trying to reduce the smaller one.

For example, if S_s is 10% and S_a is 5% the overall standard deviation is 11%

$$\rightarrow 0.11 = \sqrt{(0.10)^2 + (0.05)^2}$$

A more expensive and time-consuming analytical procedure that reduces S_a to 1% only improves S_o from 11 to 10% $\rightarrow 0.10 = \sqrt{(0.10)^2 + (0.01)^2}$

For better comprehension of the nature of uncertainty in selecting a sample for analysis, consider a random mixture of two kinds of solid particles. The theory of probability allows us to state the likelihood that a randomly drawn sample has the same composition as the bulk sample. It seems surprising to learn how large a sample is required for accurate sampling.

Suppose a mixture contains n_A of type A and n_B particles of type B. The probability of drawing A or B from the mixture is

P = probability of drawing
$$A = \frac{n_A}{n_A + n_B}$$

Q = probability of drawing
$$B = \frac{n_B}{n_A + n_B}$$

Suppose n particles are drawn at random, the expected number of particles of type A is np and the standard deviation of many drawings is known from binomial distribution to be standard deviation in sampling operation:

$$\sigma_n = \sqrt{npq} \tag{7.2}$$

7.4.1 Statistics of Drawing Particles

Example 7.1
A mixture contains 1% NaCl particles and 99% $NaNO_3$ particles. If 10^4 particles are taken, what is the expected number of NaCl particles, and what will be the standard deviation if the experiment is repeated many times?

Solution:
Expected number of NaCl particles $= np = (10^4)(0.01)$
$\quad = 100$ particles
and standard deviation $= \sqrt{npq}$
$$= \sqrt{(10^4)(0.01)(0.99)}$$
$$= 0.99$$

The standard deviation $\sqrt{npq}$ applies to both kinds of particles. The standard deviation is 99% of the expected number of NaCl particles, but only 0.1% of the expected number of $NaNO_3$ particles ($nq = 9900$).

Equation (7.2) expresses that when n particles are drawn from a mixture of two kinds of particles (such as river tissue particles and droplets of water). The sampling standard deviation will be $\sigma_n = \sqrt{npq}$, where both p and q are fraction of each particle present.

The relative standard deviation can be calculated as follows: $\dfrac{\sigma}{n} = \dfrac{\sqrt{npq}}{n} = \sqrt{\dfrac{pq}{n}}$

$$\frac{\sigma_n}{n} = \frac{\sqrt{npq}}{n} = \sqrt{\frac{pq}{n}} \tag{7.3}$$

The relative variance can be calculated as: $\left(\dfrac{\sigma_n}{n}\right)^2 = \dfrac{pq}{n} = nR^2 = pq$

$$\left(\frac{\sigma_n}{n}\right)^2 = \frac{pq}{n} = nR^2 = pq \tag{7.4}$$

Suppose that the mass of sample drawn, m, is proportional to the number of particles drawn, we can rewrite equation (7.4) in the form of sampling constant

$$nR^2 = K_s \qquad\qquad\qquad (7.5)$$

Whereby: R is relative standard deviation (expressed as percen tages) K_s is called sampling constant, K_s is the mass of sample required to reduce the relative sampling standard deviation to 1%.

Example 7.2

What mass of ^{24}Na will give a sampling standard deviation $\pm 7\%$?

Solution:

Given sampling constant equal to 36 g, i.e., $K_s = 36$ g rearrangement and substitution can be done as follows:

$$mR^2 = K_s$$

$$m = \frac{K_s}{R^2}$$

$$m = \frac{36\ g}{7^2} = 0.73\ g$$

A 0.7 g sample should given ~ 7% sampling standard deviation. Strictly speaking, this is a sampling standard deviation.

Practice 7.1

What mass of sample is expected to give a sampling standard deviation of $\pm$ 10%? Given the sampling constant $K_s = 36$.

Solution: 0.36 g

7.3.1 Dissolving Samples for Analysis

Once a bulk sample is selected, a laboratory sample must be prepared for analysis, as in Figure 7.1. A selected solid sample should be grounded and mixed so that the laboratory sample has the same composition as the bulk sample. Typically, solids are dried at 110 °C at atmospheric pressure to remove all possible adsorbed water prior to analysis.

In addition, temperature sensitive samples may simply be stored in an environment that brings them to a constant, reproducible moisture level. The entire solid should be dissolved or else it will be difficult to asure ourselves that the intended analysis is perfect. If the sample does not dissolve under mild conditions, acid digestion or fusion may be tried as an alternative.

Organic material contained in solid may be destroyed by combustion also called dry ashing or wet ashing (oxidation with liquid reagent) to place inorganic elements in appropriate form for analysis.

7.4.2   Grinding

A solid can be ground in a mortar and pestle like those shown in Figure 7.2. Materials such as ores and minerals can be crushed by stricking the pestle lightly with a hammer. The agate mortar in Figure 7.2 is designed for *grinding* small particles into a fine powder. Less expensive mortars are always more porous and more easily scratched, which leads to contamination of the sample with mortar material or the remains of previously ground samples.

It is acceptable to clean ceramic mortar by wiping with a wet tissue and washing with distilled water. Most sticky residues can be removed by grinding with the addition of 4 M HCl in the mortar or by grinding an abrasive cleaner, followed by washing with HCl and water. A boron carbide mortal and pestle is five times harder than agate and less prone to contaminate the sample.

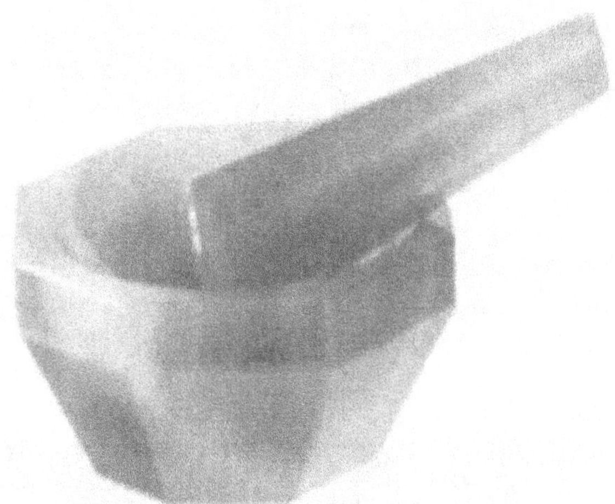

Figure 7.2: Agate Mortar

7.5   Dissolving Inorganic Materials with Acids

The nonoxidizing acids HCl, HBr, HF, H_3PO_4 dilute H_2SO_4 and dilute $HClO_4$ dissolves metals by the redox reaction.

Metals with negative reduction potentials should dissolve, although some, such as aluminum, Al, form a protective oxide coat that inhibits dissolution. Some volatile species are formed by the protonation of anions such as carbonate

,

, phosphide , fluoride

, and borate

which are likely be lost from hot acids in open vessels. Volatile metal halides such as $SnCl_4$ ad $HgCl_2$ and some molecular oxides such as OsO_4 and RuO_4 can be lost as well. Hot HF is especially useful for dissolving silicates. Glass or platinum vessels can be used for HCl, HBr, H_2SO_4, H_3PO_4, and $HClO_4$. Hydrofluoric acid, HF, should be used in Teflon, polyethylene, silver, or platinum vessels. It is recommended that the highest quality acid must always be used to minimize contamination by the concentrated reagent.

Some of substances which do not dissolve in nonoxidizing acids may dissolve in oxidizing acids such as HNO_3, hot, concentrated H_2SO_4 or hot, concentrated $HClO_4$. Nitric acid, HNO_3, attacks most metals, but not gold, Au or platinum, Pt, which dissolve in 3:1 (3 vol: HCl : 1 vol: HNO_3) called *aqua regia*.

Strong oxidant such as Cl_2 or $HClO_4$ in HCl dissolves difficult materials such as iridium, Ir, at relatively elevated temperature. A mixture of HNO_3 and HF attacks the refractory carbides, nitrides, and borides of titanium, Ti, zinc, Zn, and tungsten, W. Furthermore, hot concentrated $HClO_4$ is a dangerously powerful oxidant, whose oxidizing power increases significantly by the addition of some amounts of concentrated H_2SO_4 and catalyst such as V_2O_4 or CrO_3.

7.5.1   Dissolving Inorganic Materials by Fusion

Substances that will not dissolve in acid can usually be dissolved by a hot, molten inorganic flux. A good example can be seen in Table 7-1.

Table 7.2: Fluxes for Some Sample Dissolution

Flux	Crucible	Uses
Na_2CO_3	Pt	In dissolving silicates (clays, glass, minerals, rocks), refractory oxides, insoluble phosphates, and sulfates
NaOH or KOH	Au, Ag	Dissolves silicates and SiC. Often, frothing occurs when water is eliminated from flux; it is therefore advised to use flux first and then add the sample. It should be noted that analytical capabilities are limited by impurities in NaOH and KOH.
Na_2O_2	Zr, Ni	Strong base and powerful oxidant good for silicates which are not dissolved by Na_2CO_3. It is very useful for iron and chromium alloys.

		Because it slowly attacks the crucible, this can be avoided by coating the inside of a Ni crucible with molten Na_2CO_3, cool, and add Na_2O_2. The peroxide melts at lower temperature than the carbonate, which shields the crucible from the melt.
$K_2S_2O_7$	Porcelain SiO_2, Au, Pt	Potassium pyrosulfate ($K_2S_2O_7$) is prepared by heating $KHSO_4$ until all water is lost and foaming ceases.
B_2O_3	Pt	Very useful for oxides and silicates. A major advantage is that flux can be completely removed as volatile methyl borate [$CH_3O]_3B$) by the treatments with HCl in methanol.
$Li_2B_4O_4$ + Li_2SO_4 (2:1 wt/wt)	Pt	Good example of a powerful mixture for dissolving refractory silicates and oxides in 10-20 minutes at 1000 °C. The solidified melt dissolves very readily in 20 mL of hot 1.2 M HCl.

Fusion (melting) is carried out in Pt-Au alloy crucible at 300 °C to 1200 °C in a furnace. A disadvantage of a flux is that impurities sometimes are introduced by the large mass of solid reagent. It is good idea for the parts of an unknown to be dissolved with acid prior to fusion. Thereafter, insoluble component is dissolved with flux and the two portions are combined for analysis.

7.6   Sample Preparation Techniques

Sample preparation is the series of steps required to transform a sample so that it is suitable for analysis. Sample preparation could include dissolving the sample, extracting analyte from a complex matrix, concentrating a dilute analyte to a level that can be measured, chemically converting analyte to a detectable form, and removing or masking interfering species see chapter 6, section 6.3.2. The sample can be prepared by solid-phase extraction, preconcentration, derivatazation, and digestion of organic substance.

7.6.1   Solid – Phase Extraction

This technique uses a small volume of chromatographic stationary phase to isolate desired analytes from a sample. The versatility of solid phase extraction allows use of this technique for many purposes. The extraction removes much of the sample matrix, purification trace enrichment, desalting derivatization and class fractionation to simplify the analysis. In short, this technique uses liquid chromatography to isolate one or more analytes from a liquid mixture see details in chapter nine, section 9.3 iii.

7.6.2 Preconcentration

Trace analysis often requires preconcentration of analyte to bring it to a higher concentration prior to analysis. Metal ions in natural waters can be preconcentrated with cation-exchange resin. The latter phenomenon has been discussed thoroughly in chapter four, section 4.4.9 (ii).

7.6.3 Derivatization

Derivatization is a procedure in which analyte is chemically modified to make it easier to detect or separate. For example, formaldehyde and other aldehydes and Ketones in air, breath, or cigarette smoke can be trapped and derivatized by passing air through a tiny cartridge containing 0.35 g of silica coated with 0.3 wt% 2,4-Dinitropohenylhydrazine derivative, which is eluted with 5 mL of acetonitrile and analyzed by High Performance Liquid Chromatography (HPLC is discussed in detail in chapter nine, section 9.1 vii). The products are readily detected by their strong ultraviolet absorbance near 360 nm: see Figure 7.2.

Figure 7.3: Derivatization of Trapped Aldehyde or Ketone

7.6.4 Digestion of Organic Substances

It is a normal practice to decompose an organic compound by combustion before analysis for N, P, halogens (F, Cl, Br, I) and metals or by *digestion*. When digestion is used, a substance is first decomposed and dissolved in a reactive liquid. Frequently, sulfuric acid or a mixture of H_2SO_4 and HNO_3 is added to the organic compound and gently heated in a microwave bom for about 10 to 20 minutes. This will ensure all particles have completely dissolved and the solution has a uniform black appearance. After the latter has cooled to room temperature, the dark colour is destroyed by addition of hydrogen peroxide (H_2O_2) or HNO_3, and heat. It is recommended to analyze a decomposed sample after digestion.

7.7 Problems

7.7.1. A soil sample contains some acid-soluble inorganic matter, some organic material, and some minerals that do not dissolve in any combination of hot acids that you try. Outline the procedure of which you are going to use to dissolve entire sample.

7.7.2. Suppose a container contains 60, 000 red balls and 440, 000 blue balls.
(a) If you draw a random sample of 1000 balls from the container, what are the most probable numbers of red and blue balls?
(b) Repeat the experiment by putting back the balls in the container. What will be the absolute and relative standard deviation for the numbers in (a) after many drawings of 1000 balls?
(c) What sample size is required to reduce the sampling standard deviation of red balls to $\pm 2\%$?

7.7.3. In a given experiment the sampling constant was found to be $K_s = 20$ g.
(a) What mass of sample is required for a $\pm 2\%$ sampling standard deviation?
(b) How many samples of the size in (a) are required to produce 90% confidence that mean is known to be within 1.5%?

7.74. Explain what is meant by the statement "Unless the complete history of any sample is well known, a chemist should not to spend time in analyzing it."

7.7.5. Explain why it is important to be extra cautious when we are planning to store a sample.

7.7.6. Why does aluminium metal tend to be resistant to dissolving by the redox reaction despite it having negative reduction potentials?

7.7.7. Outline possible steps you are going to take for substances which are not soluble in acid when you are planning to prepare samples for analysis.

7.7.8. What is the difference between a random heterogeneous material and a segregated heterogeneous material?

7.7.9. What is the difference between a random sample and composite sample?

7.7.10. With examples, explain what it means by derivatization.

7.8 References

Chai, A. and Kuet, L. (2004). Solid Phase Extraction Cleanup for the Determination of Organochlorine Pesticides in Vegetables, *Malaysia J. Chem. 6*, 39.

Demeestere, K., Dewulf, J., Wittle, B. and Langenhove, H. V. (2007). Sample Preparation for the Analysis of Volatile Organic Compounds in Air and Water Matrices, *J. Chromatogra. A, 153*, 130.

Guy, R. D., Ramaley, L. and Wentzel, P. D. (1998). "Experiment in the Sampling of Solids for Chemical Analysis," *J. Chem. Ed. 75*, 1028.

Hartman, J. R. (2000). "In-Class Experiment on the Importance of Sampling Techniques and Statistical Analysis of data," *J. Chem. Ed. 77*, 1077.

Harvey, D. (2002). "Two Sampling in Quantitative Chemical Analysis," *J. Chem. Ed. 79*, 360.

Ras, M. R., Borrull, F. and Marce, M. (2009). Sampling and Preconcentration Techniques for Determination of Volatile Organic Compound in Air Samples. *TrAc Trends in Analytical Chemistry, 28*, 347.

Rawa-Adkonis, M. Wolska, L. and Namieshik, J. (2003). Mordern Techniques of Organic from Environmental Matrices, *Critc. Rev. Anal. Chem, 33*, 199.

Sawyer, D. T. Sobkowiak, A. and Roberts Jr., J. L. (1995). *Electrochemistry for Chemists*, 2nd ed. Wiley New York.

Vitt, J. E. and Engstrom, R.C. (1999). "Effect of Sampling Size on Sampling Error," *J. Chem. Ed. 76*, 99.

Westborn, R., Thoneby, L., Zorita, S. L. Mathiasson, J. Bjorklund, J. (2004). Development of a Solid-phase Extraction Method for the Determination of Organochlorinated Biphenyl in Water, *J. Chromatogr. A, 1033*, 1.

Chapter 8

THE CHEMISTRY OF ACIDS AND BASES

8.1   Introduction

The reactions of acids and bases cannot be over-emphasized because such reactions are one of the most important types of reactions. In this chapter we will look at the Arrhenius, Bronsted-Lowery, and Lewis theories of acids and bases, but special focus will be on acid-base titrations. However, the area of acid-base chemistry is an area in which most students commonly have difficulties, so students should spend extra time to master these important concepts.

8.2   Acids-Bases and Arrhenius

The word '*acid*' comes from the Latin word *acidus*, meaning sour. Acids are characterized by their sour taste, their ability to be corrosive, redden blue vegetable colours and lose all these properties when reacted with alkalies (bases). The term 'alikali' was derived from the Arabic word for the ashes that come from burning certain plants. Potash (potassium carbonate) is one of the products formed by this process, and due to the fact that water solutions of potash feel soapy and taste bitter, the term was commonly used for other substances having those properties. It was also realized that acids react with alkalies to form salts. Thus, alkalies were considered to be the '*bases*' for these salts, and the term 'base' became used to describe alkalies.

Bases are characterized by their soapy/slippery solutions, ability to restore vegetable colours reddened by acids and ability to react with acids to form salts. The term salt means any ionic compound whose cation could come from a base and whose anion could come from an acid in an acid-base reaction.

8.2.1   Arrhenius Definition of Acids and Bases

Arrhenius defined an *acid* as a substance that contains hydrogen and produces hydrogen (H^+) by ionic dissociation in water (expression 8.1).

On the other hand, a *base* is a substance that produces hydroxide (OH^-) in water (expression 8.2-8.4).

141

The reaction of acid and base produces a salt and water (expression 8.5-8.6).

Arrhenius's concept is limited to aqueous solutions, because it refers to ions (H^+ and OH^-) derived from water.

8.3   The Hydronium Ion and Water Autoionization

A H^+ from water is often represented as *hydronium* ion (H_3O^+), as described in chapter two, section 2.6. Two water molecules can react with each other to produce a hydronium ion and a hydroxide ion by proton transfer from one water molecule to the other in what is called *autoionization* reaction: expression 8.7.

8.4   The Bronsted-Lowery Concept of Acids and Bases

Bronsted-Lowery defined an *acid* as any substance that can donate a proton to another substance: this means an acid can be neutral molecules, cations or anions.

Conversely, a *base* is any substance that can accept a proton from another substance; therefore, bases can be neutral molecules or anions.

$$NH_3\ (aq) \quad + \quad H_2O\ (l) \quad \rightleftharpoons \quad NH_4^+\ (aq) \quad + \quad OH^-\ (aq) \tag{8.11}$$

$$PO_4^{3-}\ (aq) \quad + \quad H_2O\ (l) \quad \rightleftharpoons \quad HPO_4^{2-}\ (aq) \quad + \quad OH^-\ (aq) \tag{8.12}$$

Acids capable of donating only one proton are called *monoprotic* acids, whereas acids capable of donating two or more protons are called *polyprotic* acids. Anions of polyprotic acids having charges of two negatives (2-) or more can accept more than one proton, and therefore act as polyprotic bases. The following are common examples: CO_3^{2-}, $C_2O_4^{2-}$, S^{2-}, SO_4^{2-}, HPO_4^- and PO_4^{3-}.

Molecules and ions that can behave as either a Bronsted-Lowery acid or base are called *amphiprotic* substance. One of the most common amphiprotic substances is *water*. Water acts as a base in reactions with acids, such as expressions 8.8, 8.9, 8.10, as an acid in its reactions with base, such as expressions 8.11, 8.12, and as both acid and base in its autoionization reaction given in expression 8.7. In addition, there are other important amphiprotic substances with intermediate forms of polyprotic acids that have not yet had all ionizable hydrogen atoms removed, such as HS^-, $H_2PO_4^-$ and HPO_4^-.

8.4.1 Conjugate Acid-base Pair

Those substances that differ by an H^+ unit in their chemical formulas are called a *conjugate acid-base pair*. This means CO_3^{2-} is the *conjugate base* of HCO_3^-, and that HCO_3^- is the conjugate acid of CO_3^{2-}. Nevertheless, HCO_3^- is amphiprotic; thus, it is also the conjugate base of H_2CO_3, and H_2CO_3 is the conjugate acid of HCO_3^-. Below is the general form of a Bronsted-Lowery acid-base reaction:

where Acid 1 and Base 1 are conjugate acid-base pair and Base 2 and Acid 2 are conjugate acid-base pair. A good example is clearly shown in the reaction in expression 8.8, HNO_3 and NO_3^- is a conjugate acid-base pair, and H_2O and H_3O^+ is conjugate acid-base pair. Also in a similar reaction expression 8.11, NH_3 and NH_4^+ is a conjugate acid-base pair, and H_2O and OH^- is a conjugate acid-base pair.

8.5   Relative Strength of Acids and Bases

There are some variations among acids in terms of proton donation H^+: some acids are better proton donors in water than other acids, and some bases are better proton acceptor than others. Take an example of a dilute nitric acid which is almost 100% ionized, and because of that tendency nitric acid is considered to be a strong acid. This implies that a 0.1 M aqueous solution of HNO_3 in fact consists of 0.1 M H_3O^+ and 0.1 M NO_3^- and few, if any, molecules of HNO_3. Conversely, organic acid like acetic acid ionizes very slightly such that a 0.1 M aqueous solution of acetic acid consists of almost 0.001 M H_3O^+, 0.001 M $CH_3CO_2^-$ and 0.099 M CH_3CO_2H. This means that only about 1% of molecules ionize in a 0.1 M CH_3CO_2H solution, and acetic acid is a well known acid. Converseely, oxide ion is too strong a base to exist in aqueous solution. It reacts completely with water to produce hydroxide ions.

So the dissolved lithium oxide in water can be expressed as in expression 8.15

Conversely, aqueous ammonia and the carbonate ion produce only a very small concentration of hydroxide ions and due to that they are known as being weak Bronsted bases.

$$NH_3 \text{ (aq)} \quad + \quad H_2O \text{ (l)} \quad \rightleftharpoons \quad NH_4^+ \text{ (aq)} \quad + \quad OH^- \text{ (aq)} \qquad (8.16)$$

$$CO_3^{2-} \text{ (aq)} \quad + \quad H_2O \text{ (l)} \quad \rightleftharpoons \quad HCO_3^- \text{ (aq)} \quad + \quad OH^- \text{ (aq)} \qquad (8.17)$$

Often, strong acids have very weak conjugate bases. A good example is aqueous HCl, which is considered to be a strong acid because it donates a proton to water to form its conjugate base Cl⁻; this means Cl⁻ must have little tendency to retain the proton. Generally speaking, the stronger the acid, the weaker its conjugate base, so it can be summarized as follows:

- Strong acids have very weak conjugate bases.
- Weak acids have weak conjugate bases, and
- Very weak acids have strong conjugate bases.

On the contrary, the stronger the base the weaker it is conjugate acid. Thus, the existing differences between the reaction of HNO_3 and CH_3CO_2H with water illustrate a very important principle in Bronsted acid-base theory. This means all proton transfer reactions proceed largely from the stronger acid-base pair to the weaker acid-base pair.

8.6   Strong Acids and Bases

In fact, hydronium ion is actually the strongest acid that can exist in water. This is considered as a leveling effect of water, because the proton of these acids is brought to the level of H_3O^+. It should clearly be remembered that this applies to the removal of the first proton of H_2SO_4, since HSO_4^- is a negatively charged species that is a weaker acid than H_3O^+.

It is apparent that the hydroxide ion is the strongest base known that can exist in water. Most bases will react with water to form OH^- and their respective conjugate acids and therefore turn out to be with the same strength in water. This means that they are brought to the level of OH^- in water. Take an example of hydride H⁻ ion which is too strong base to exist in water and the vigorous reaction of H⁻ with H_2O is as follows in expression 8.18:

Typically, the most common examples of strong base are the soluble metal oxides and hydroxides of elements of Group IA and IIB.

8.7   Weak Acids and Bases

It should be clearly noted that the majority of acids and bases are weak. Indeed, the relative strength of an acid or base can be expressed quantitatively with equilibrium constant. This can be easily expressed by a general weak acid HA, as follows in expression 8.19 and 8.20:

$$K_a = \frac{[H_3O^+][A^-]}{[HA]} \tag{8.20}$$

whereas K has subscript 'a' which indicates that it is an equilibrium constant for a weak acid in water and the concentration of water has been neglected from the equilibrium in keeping with the principles developed in chapter two. If the reaction is *reactant* favoured, it means the value of Ka is less than 1, which indicates the product of the equilibrium concentration of the hydronium ion and the equilibrium concentration of the conjugate base of the acid is less than the equilibrium concentration of acid.

The general expression for weak base can be written as follows in expressions 8.21 and 8.22:

$$K_b = \frac{[BH^+][OH^-]}{[B]} \tag{8.22}$$

whereas K has subscript 'b' which indicates that it is an equilibrium constant of weak base in water and the concentration of water has been neglected from the equilibrium in keeping with the principles developed in chapter two. Also, if the reaction is a *reactant* favoured, the value of K_b is less than 1. It should be recognized that the stronger the acid the weaker is its conjugate base. Table 8.1 shows a classification of acids and bases on the basis of their respective K_a and K_b.

Table 8.1: Classification of Acids and Bases based on K_a and K_b

Acid Strength	K_a	Conjugate Base Strength	K_b
Strong	> 1	Very weak	< 10^{-16}
Weak	1 to 10^{-16}	Weak	1 to 10^{-16}
Very weak	< 10^{-16}	Strong	> 1

Cations as weak acids:

A good example is the ammonium ion, NH_4^+, which is the conjugate acid of the weak base ammonia, NH_3 and therefore, acts as a weak acid.

On the other hand, hydrated metal cations of both high charge and small size act as weak acids see expression 8.24. A good example, $[Al(H_2O)_6]^+$, is almost as strong an acid as CH_3CO_2H, which can endanger the environment, because acid rain can leach Al^{3+} ions from the soil, as follows:

Anions as weak acids:
Of all these weak acids, hydrogen sulfate ion is the strongest:

Neutral molecules as weak bases:
Good and common examples of such molecules include ammonia. The latter is also the parent compound of a large series of compounds called amines. Often, amines are formed by substitution of hydrogen atoms of ammonia by different alkyl groups, for example methyl groups in methylamine, CH_3NH_2; dimethylamine, $(CH_3)_2NH$; and trimethylamine, $(CH_3)_3N$. An important thing to be noted from these compounds is their ability to act as weak base by using a lone pair of electrons on nitrogen atom that can accept a proton to form a coordinate covalent bond with the H^+.

Example 8.1
The concentrations of H_3O^+ ion in a sample of lemon juice is 2.5 x 10^{-3} M. Calculate the concentration of OH^- ions, and classify the solution as acidic or basic.

Solution:
When $[H_3O^+]$ is well known, the OH^- concentration can be calculated from the expression

$$[OH^-] = \frac{K_w}{[H_3O^+]}$$

$$[OH^-] = \frac{K_w}{[H_3O^+]} = \frac{1.0 \ x \ 10^{-14}}{2.5 \ x \ 10^{-3}} = 4.0 \ x \ 10^{-12}$$

Due to the fact $[H_3O^+] > [OH^-]$, the solution is acidic.

8.8 Water and the pH Scale

It is apparent that the stronger the Bronsted acid or base, the larger the concentration of H_3O^+ or OH^- for a given concentration of acid or base, respectively. Clear understanding of pH (see section 2.6) enables us to measure these concentrations quantitatively and compare both acid and base strengths.

Water ionization constant, K_w
Autoinization of water produces both hydronium and hydroxide ions. Usually, the equilibrium position for autoionization reaction in equation 8.26 lies far to the left, the reason being H_3O^+ is a much stronger acid than H_2O and OH^- is a much stronger base than H_2O.

It has been noted from the experiment that only 2 out of approximately 10^9 (a billion) water molecules are ionized at 25 ^{o}C. This can quantitatively be expressed in terms of the equilibrium as follows:

$$K = \frac{[H_3O^+]\,[OH^-]}{[H_2O]^2} \tag{8.27}$$

In actual fact, when the molar concentration of pure H_2O is 55.5 M at room temperature, it does not matter how much water is present, and the concentration of H_2O can be substituted into the equilibrium constant giving $K\,[H_2O]^2 = [H_3O^+]\,[OH^-]$, which is applicable in both aqueous solutions and pure water, $K_w = [H_3O^+]\,[OH^-]$. The constant K_w is known as the ionization constant of water. In the case of pure water, both H_3O^+ and OH^- are formed in equal amounts through *autoionization* of water. Typically, a very careful electrical conductivity measurement of pure water often shows that:
$[H_3O^+] = [OH^-] = 1.0 \; x \; 10^{-7}$ M at 25 ^{o}C, so

$$K_w = [H_3O^+]\,[OH^-] = (1.0 \; x \; 10^{-7})\,(1.0 \; x \; 10^{-7}) = 1.0 \; x \; 10^{-14} \tag{8.28}$$

at 25 ^{o}C. Due to that reason hydronium and hydroxide concentrations in pure water are both $1.0 \; x \; 10^{-7}$ M at 25 ^{o}C, and it is because of that water is considered to be *neutral*. Indeed, it should be clearly noted that K_w is a *temperature dependent* constant, thus the concentration of H_3O^+ and OH^- found in pure water and other neutral solutions at other temperatures vary with temperature.

Obviously, the addition of acid or base in water will raise the concentration of one and lower the other. In opposing this addition, Le Chatelier's principle predicts H_3O^+ will react with OH^- to form water in the process and lower the $[OH^-]$ until the product $[H_3O^+]\,[OH^-]$ equals to $1.0 \; x \; 10^{-14}$ at 25 ^{o}C. On the other hand, adding base to water raises $[OH^-]$, and Le Chatelier's principle predicts that OH^- will react with H_3O^+ to form water and lower the $[H_3O^+]$ until the product $[H_3O^+]\,[OH^-]$ again equals to $1.0 \; x \; 10^{-14}$ at 25 ^{o}C. Therefore, hereunder are summaries for aqueous solution:

- In a neutral solution, $[H_3O^+] = [OH^-]$ and both are equal to $1.0 \; x \; 10^{-7}$ M.
- In an acidic solution, $[H_3O^+] > [OH^-]$. $[H_3O^+] > 1.0 \; x \; 10^{-7}$ M and $[OH^-] < 1.0 \; x \; 10^{-7}$ M, and
- In a basic solution, $[H_3O^+] < [OH^-]$. $[H_3O^+] < 1.0 \; x \; 10^{-7}$ M and $[OH^-] > 1.0 \; x \; 10^{-7}$ M.

8.9   Polyprotic Acids and Bases

Polyprotic acids often ionize in successive steps, a good example being phosphoric acid. The following are phosphoric acid ionization steps (8.29, 8.30, and 8.31) and their respective K_a values:

$$H_3PO_4 \text{ (aq)} + H_2O \text{ (l)} \rightleftharpoons H_3O^+ \text{ (aq)} + H_2PO_4^- \text{ (aq)} \quad K_{a1} = 7.5 \times 10^{-5} \quad (8.29)$$

$$H_2PO_4^- \text{ (aq)} + H_2O \text{ (l)} \rightleftharpoons H_3O^+ \text{ (aq)} + HPO_4^{2-} \text{(aq)} \quad K_{a2} = 6.2 \times 10^{-8} \quad (8.30)$$

$$HPO_4^{2-} \text{ (aq)} + H_2O \text{ (l)} \rightleftharpoons H_3O^+ \text{ (aq)} + PO_4^{3-} \text{ (aq)} \quad K_{a3} = 3.6 \times 10^{-13} \quad (8.31)$$

You can carefully study how the successive K_a values get smaller, due to the fact that it becomes to some extent difficult to remove H^+ to the more negative on the acidic specie. In actual fact, every successive K_a is less than the previous one by a factor ranging between 10^4 to 10^6, and this is a typical example of *oxoacids* which are composed of H, O, and nonmetal. Indeed, this also means that every successive ionization step produces about a million times less hydronium ion than the step before it; thus, the pH of the solution primarily depends on the $[H_3O^+]$ generated by the first ionization step.

Polyprotic acids principles apply as well to their respective conjugate anions. This means:

- Every successive K_b is a factor of 10^4 to 10^6 less than the one before it, because the less negative the anion the lower the attraction for a H^+ ion from water, and

- Both pH and pOH depend basically on the concentration of OH- generated by the first step.

8.10 Molecular Structure, Bonding and Acid-Base Behaviour

Let us consider the dissociation of an acid HA, which occurs in steps so that we can examine the correlation between molecular structure, chemical bonding and acid strength. Those steps are as follows:

(i) When H-A bond breaks, each atom retains the electron it brought to the bond.

(ii) Loss of an electron by H to form H^+.

(iii) Gain the electron by A to form A-.

(iv) The hydration of H^+ and A- ions results in the formation of aqueous ions.

It is possible to ignore step (8.33) because it is common to all acid dissociations, and as well ignore step (8.35) due to the fact that hydration has almost the same effect for analogous cases. Now we can focus on the H-A, bond strength and the fragment *A electron affinity* when analyzing the strengths of similar acids. Generally speaking the weaker the *H-A* bond and the higher the electron affinity of the fragment *A*, the stronger the acid. Before we proceed, let us define what *electron affinity* means: this is the energy change that occurs when an electron is added to a gaseous atom (see expression 8.36):

$$A(g) \; + \; e^- \longrightarrow A^-(g) \tag{8.36}$$

It should always be noted that in the above H-A dissociation factors act in opposing directions, as in the case of hydrogen halides.

For the case of hydrogen halides, bond strength decreases with increasing size of *A* from Fluorine, F, to Iodine, I, whereas the electron affinities of halogen atom increase from F to Cl and thereafter decrease with increasing size from chlorine, Cl, to I. Thus, based on electron affinity concept regarding **A**, HCl should be the strongest HA type of acid. Furthermore, according to bond strength concept, HI should be the strongest HA acid in the series. Nevertheless, acid strength from HF to HI shows regular increase that parallels the regular decrease in H-A bond strength and indicates that the decrease in bond strength is the controlling factor in this series. Indeed, a similar pattern exists for binary hydrides of elements of other groups, which means that acid strength increases from H_2O to H_2Te in group VIA. However, it should be recognized that the combined effects of electron affinity of fragment **A** and hydration energy for fragment **A** dominate as you go across a period; this means acidity increases in the order CH_4, NH_3, H_2O and HF for second period nonmetal binary hydrides.

On the other hand, *oxoacids* which is composed of H, O, and a nonmetal, the hydrogen atom that is ionized, is often bonded to oxygen, O, atom, and the acidity is related to the composition of the rest of the molecule. The acidity strength trend is as follows: HOCl, $HClO_2$, $HClO_3$ and $HClO_4$, due to the increase in the number of O atoms attached to Cl that withdraws electron density from Cl and results in a shift in electron density in the Cl-O-H bonds towards Cl through an *inductive effect* (attraction of electrons from adjacent bonds caused by a more electronegative atom). The latter makes the H atom positive or more protonic, and in turn the acid gets stronger, so $HClO_3$ having two O atoms is expected to be stronger than H_3PO_4 which has only one O atom. In fact, all oxiacids having two or more O atoms attached to the central atom often are expected to be stronger acids.

For the case of $HBrO_3$ and $HClO_3$, involving central atoms from the same group and the same number of O atoms, the higher electronegativity of Cl causes the H to be more positive and $HClO_3$ to be the strongest acid. Generally speaking, the more electronegative the central atom, the stronger the acid when one is comparing with acids of elements of the same group having the same number of O atoms.

It is good to note that the H bonded directly to O in acetic acid, CH_3CO_2H, is ionizable. The H attached directly to O atom is more positively charged, more protonic and O is good at accommodating the negative charge left when the O-H bond breaks than C would be, if a C-H bond would break. Typically, this is a general behavior of carboxylic acids.

8.11   Lewis Concept of Acids and Bases

Basically, the Lewis acid-base theory involves the sharing of an electron pair between an acid and a base, as opposed to the transfer of a proton. In fact, *Lewis acid* is a substance that can accept a pair electron from another atom to form a new bond. *Lewis base* is a substance that contains atom with a lone pair of electrons that can be donated to another atom to form a new bond. This can be seen in the expression 8.37 as:

where A represents the Lewis acid, and B represents the Lewis base and AB represents the product, which often is called *adduct* or *complex*. The bond which exists between Lewis acid and Lewis base is a *coordinate covalent* bond. The Lewis acid supplies an empty orbital that overlaps with a filled orbital from the Lewis base. The Lewis acid theory, however, is more general than the Bronsted theory. This implies that every Bronsted acid is also a Lewis acid and that every Bronsted base is a Lewis base. Conversely, every Lewis base is potentially a Bronsted base, because it has a lone pair of electrons of which it uses to form a bond with H^+, but not every Lewis acid is a Bronsted acid. It should be clear that not every Lewis acid contains hydrogen atom.

8.11.1 Cations as Lewis Acid

Obviously, all metal cations for example, Al^{3+}, Fe^{3+}, Zn^{2+}, Mg^{2+}, Na^+, and Ag^+ are electron deficient, which is why they are considered as potential Lewis acids. Thus, the formation of hydrated metal cations is a vivid example of a Lewis acid-base reaction. The bonding in hydrated metal cations often explains the reason why they form acidic solution. Generally, most of the metal ions also act as Lewis acids in forming complexes with Lewis base ammonia. Indeed, the *amphoteric* characteristic of metal hydroxide involves the metal hydroxide acting separately as a Lewis acid and a Bronsted base. For example, consider $Al(OH)_3$

- in its reaction with bases, $Al(OH)_3$, acts as a Lewis acid:

- in its reaction with acids, $Al(OH)_3$, acts as Bronsted base:

$$Al(OH)_3 \text{ (s)} \quad + \quad 3\,H_3O^+ \text{ (aq)} \longrightarrow Al^{3+} \text{ (aq)} \quad + \quad 6\,H_2O \text{ (l)} \qquad (8.39)$$

Lewis acid theory has provided clear explaination for the behavior of oxides of nonmetals as an acid. The most electronegative oxygen atoms withdraw electrons from the central atom, causing it to be partial positively charged and making it susceptible to being attacked by Lewis bases, for example, OH^- or H_2O.

8.12   Identifying Strong and Weak Acids and Bases

Due to the fact that there are only a few strong acids and bases, differentiating weak and strong acids and bases can be done through memorization of a list of strong acids and bases. These can be seen in Table 8.2 hereunder.

Table 8.2: List of Some Strong Acids and Strong Bases

Strong Acids	
Nitric acid	HNO_3
Sulfuric acid	H_2SO_4 (for loss of first H^+ only)
Halic acid	$HClO_3$, $HBrO_3$
Perhalic acids	$HClO_4$, $HBrO_4$
Hydrohalic acids	HCl, HBr, and HI
Strong Bases	
Group I A hydroxides	LiOH, NaOH, KOH, RbOH, CsOH
Soluble Group 2 A hydroxides	$Ca(OH)_2$, $Sr(OH)_2$, $Ba(OH)_2$
Stronger than OH^-	S^{2-}, $C_2H_5O^-$, NH_2^-, H^-, CH_3^-

8.13   Acid-Base Titration Curves

Acid-base titration curves are often the plot of pH versus mL (acid or base) added from a burette or autotitration set-up. This is essentially used to determine the equivalence point of the reaction or to select an appropriate acid-base indicator for the reaction. Indeed, it is important to note that there are differences between the titration curves for the titration of a strong acid with a strong base, and titration of a weak acid with a strong base. These are as follows:

- It is always the case for strong acid solution pH to be lower than that of the weak acid of equal concentrations, because the strong acid ionizes 100%, thus yielding a higher $[H_3O^+]$ than weak acid which ionizes just slightly.
- Strong acid pH increases steadily at first, whereas the pH of weak acid solution shows an initial jump, which is followed by a buffer region in which changes happen relatively gradually.

- At half-neutralization point of weak acid the pH is equal to the pKa of the acid, because one half of the acid has been converted to its conjugate base and the ratio is:
$$1 = \frac{[conjugat\ base]}{[acid]}$$

- There is no real corresponding point during titration of strong acid with a strong base due to the existence of little, if any, buffering action in such a system.

- At equivalence point the pH of strong acid with strong base reaction is seven, and conversely, that of weak acid with strong base reaction is > 7 because of the hydrolysis of conjugate weak base anion that is often formed.

- The observed pH rise near the equivalence point is relatively larger for strong acid than it is for weak acid. Generally speaking, the stronger the acid, the higher the rise in pH at equivalence point. However, for the case of weak acid with weak base titrations, there is a buffer region before and after the equivalence point; thus, equivalence point cannot be determined using ordinary procedures.

At half-neutralization point the pOH for the titration of a weak base with strong acid is similarly equal to the pK_b of the base; thus, titration curves can also be used to determine K_a and K_b values.

8.14   Calculations for Titration Curves

Typically, there are four regions which require attention along the titration curves: the starting point, the region between the starting point and the equivalence point, the equivalence point and the region beyond the equivalence. In fact, we will consider the procedure for calculating each of these regions, particularly for strong acid-strong base, weak acid-strong base, weak base-strong acid titrations, and no calculation will be attempted for weak diprotic acid-strong base due to its complexity.

8.14.1 Titration of Strong Acid-Strong Base

If one can make a good follow-up of what is going on in the titration curve, then he/she will be able to understand what is happening during the titration, and make an interpretation of the titration curve with the help of the following concepts hereunder:

- At starting point (0.00 mL base added) pH is due to [H_3O^+] formed by 100% ionization of the strong acid.

- The pH at any point between the starting point and the equivalent point is due to excess [H_3O^+], which can be calculated by using expression 8.40.

$$[H_3O^+] = \frac{original\ moles\ of\ acid\ -\ moles\ of\ base\ added}{volume\ of\ acid\ (mL) +\ volume\ of\ base\ added} \tag{8.40}$$

whereas mol acid = mL of acid x M acid; and mol of base = mL base x M base.

- At the equivalence point pH is 7 because the cation of the salt formed comes from a strong base and obviously does not react significantly with water, and anion of the salt formed comes from a strong acid and does not react sufficiently with water.
- At any point pH beyond the equivalence point is due to [OH-] formed by 100% dissociation of excess strong base, which can be calculated by using expression 8.41.

$$[OH^-] = \frac{moles\ of\ base\ added - original\ moles\ of\ acid}{volume\ of\ acid\ (mL) + volume\ of\ base\ added} \qquad (8.41)$$

Having the above concept in mind, you can analyze the titration reaction equation as follows:

Step 1

Write the balanced chemical equation between titrant and analyte.

Step 2

Use the reaction equation to calculate the composition, and pH after each addition of titrant. Consider the titration of 50 mL of 0.020 M KOH with 0.100 M HBr.

Reaction equation is

$$K = \frac{1}{Kw} = \frac{1}{10^{-14}} = 10^{14}$$

From the above relationship between K and K_w, whereas $K = 10^{14}$ it makes sense to say that it "goes to completion" prior to the equivalence point, and any amount of H^+ added will consume a stoichiometric amount of OH^-. Consider if at the starting point you have to calculate the volume of HBr (V_e) needed to reach the equivalence point. At the starting point the volume of KBr (V_e) needed to reach the equivalence point can be calculated as follows:

V_e (mL) x (0.100 M) = (50.00 mL) x (0.0200 M)

$$V_e = \frac{(50.00\ mL)\ x\ (0.0200\ M)}{(0.100\ M)} = 10.00\ mL$$

V_e = 10.00 mL.

After 10.00 mL of HBr have been added, the titration is complete. Prior to V_e, excess, unreacted OH^- is present in the solution. After V_e there is excess H^+ in solution. In the titration of a strong base with a strong acid, three important regions of the titration curve require different approaches of calculations:

- Before the equivalence point, the pH is determined by excess amount of OH^- in the solution.

- At the equivalent point, added H⁺ is just sufficient to react with all OH⁻ to make H_2O. The pH is determined by the dissociation of H_2O.
- After the equivalence point, pH is determined by excess H⁺ in the solution. Now let us look at one sample calculation for each region in example 8.1.

Example 8.1

Region 1: Before the Equivalence Point
Before addition of HBr titrant from the burette, the flask of analyte contains 50.00 mL of 0.0200 M KOH, which amounts to (50.00 mL) x (0.0200 M) = 1.00 mmol of 0.0200 M KOH = 1.00 mmol of OH⁻. The following relationship hereunder should be clearly understood when dealing with such problem.

$$\frac{mmol}{mL} = \frac{mol}{L} = M \text{ and}$$

$$mmol = \frac{mol}{L} \; x \; mL$$

Suppose we have just added 3.00 mL of HBr, this means we are adding 3.00 mL x 0.100 M

= 0.300 mmol of H⁺ which consumes 0.300 mmol of OH⁻.

The total volume in the flask is now 50.00 mL + 3.00 mL = 53.00 mL. Therefore, the concentration of OH⁻ in the flask is

$$[OH^-] = \frac{0.700 \; mmol}{53.00 \; mL} = 0.0132 \; M$$

Having the concentration of OH⁻, it should be straightforward to calculate the pH at this point.

$$[H^+] = \frac{K_w}{[OH^-]} = \frac{1 \; x \; 1.0 \; x \; 10^{-14}}{0.0132} = 7.58 \; x \; 10^{-13} \; M$$

$$[H^+] = 7.58 \; x \; 10^{-13} \; M \; \rightarrow pH = 12.12$$

It should be noted that the volume of acid added is designated V_a and pH is expressed to the 0.01 decimal place, regardless of what is justified by significant figures. This is due to consistency reasons, and 0.01 is near the limit of accuracy in pH measurements.

Region 2: At the Equivalence Point
At the equivalence point, enough H$^+$ has been added to react with all the OH$^-$. We could prepare the same solution by dissolving KBr in water. The pH is determined by the dissociation of water:

$$K_w = x^2 = 1.0 \times 10^{-14} \Rightarrow x = 1.0 \times 10^{-7} \; M$$
pH = 7.00

The pH at the equivalence point in the titration of any strong base (or acid) with strong acid (or base) Figure 8.1 will be 7.00 at 25 $^\circ$C. In fact, you will soon discover that the pH is not 7.00 at the equivalence point in the titration of weak acids or bases. The pH 7 is 7.00 only if the titrant and analyte are both strong.

Region 3: After the Equivalence Point
Beyond the equivalence point, excess HBr is present. For example, at the point where 10.50 mL of HBr have been added, there is an excess of 10.50 - 10.00 = 0.50 mL.
Excess [H$^+$] = (0.50 mL) (0.100 M) = 0.050 mmol
Using the total volume of solution (50.00 + 10.50) mL = 60.50 mL), we should be able to find the pH.

$$[H^+] = \frac{0.050 \; mmole}{60.50 \; mL} = 8.26 \times 10^{-4} \; M$$
[pH = -log (8.26 x10^{-4}) = 3.08

The Titration Curve
The titration below is a graph of (pH versus V_a) the volume of acid added. The abrupt change in pH near the equivalence point is a characteristic of all analytically useful titrations. The curve is steepest at the equivalence point, which means that the slope is greatest. The pH at equivalence point is only 7.00 in a strong acid-strong base. If one or both of the reactants are weak, the equivalence point pH is not 7.00.

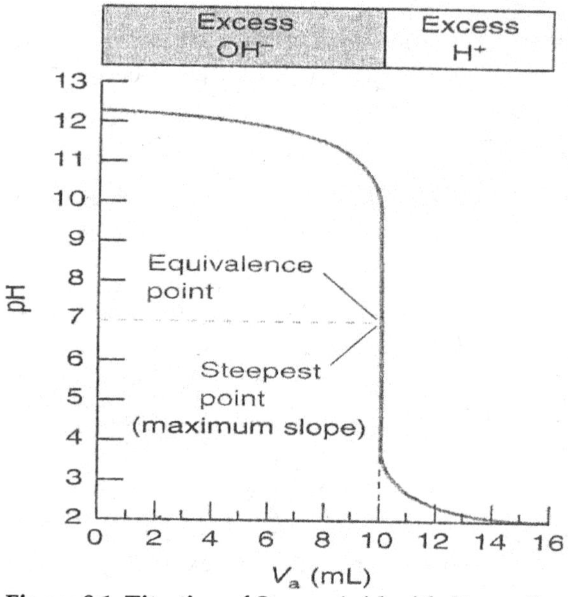

Figure 8.1: Titration of Strong Acid with Strong Base

Practice 8.1

Find the pH when 12.74 mL of 0.08742 M NaOH have been added to 25.00 mL of 0.0666 M $HClO_4$.

 Solution: pH = 1.83

8.14.2 Titration of Weak Acid with Strong Base

The most important concepts that should be put under consideration before determination of pH at different points along the titration curve are as follows:

- At starting point (0.00 mL base added) pH is due to $[H_3O^+]$ formed by ionization of the weak acid.

- It should be clearly noted that pH at any point is due to the ratio of HA/A⁻ buffer that is created as some of the weak acid, HA, is converted to its conjugate base A⁻, by the reaction with the strong base. Often, pH can be calculated through substitution of the moles of acid remaining and the conjugate base into either expressions (8.43 or 8.44).

$$[H_3O^+] = \frac{[\,acid\,]}{[conjugate\ base]} \cdot Ka \tag{8.43}$$

$$pH = pKa + log\left(\frac{[conjugate\ base]}{[acid]}\right) \tag{8.44}$$

These quantities can be calculated by using,

mol acid remaining = original mol acid − mol base added, and $\qquad$ (8.45)

mol conjugate base formed = mol base added   (8.46)
where mol acid = $mL_{acid} \cdot M_{acid}$; and mol base = $mL_{base} \cdot M_{base}$

- The pH at equivalence point is in fact the result of hydrolysis of conjugate weak base anion, A- that is formed during the titration.

- The pH measured at any point beyond equivalence point is due to [OH-] formed by 100% dissociation of the excess strong base. This can similarly be calculated using expression (8.41).

Having the above concept in mind, you can look at the titration reaction equation, and systematically do the following:

Example 8.2
Consider the titration of 50.00 mL of 0.0200 M MES with 0.100 M NaOH. MES is an abbreviation for 2-(N-morpholino) ethane sulfonic acid, which is a weak acid with pKa = 6.27. It is commonly used in Biochemistry as a buffer for the pH 6 region.

$$\overset{+}{N}HCH_2CH_2SO_3^- + OH^- \longrightarrow NCH_2CH_2SO_3^- + H_2O \qquad (8.48)$$

$$HA \qquad\qquad A^-$$

MES   pKa = 6.27

The reaction is the reverse of the K_b reaction for the base A-. The equilibrium constant is

$$\frac{1}{K_b} = \frac{1}{K_w \big/ K_{HA}} = 5.4 \times 10^7$$

The equilibrium constant is so large that we can say the reaction goes "to completion" after each addition of OH-. The volume needed to reach the equivalence point can be calculated based on what we know, that is, 1 mole of OH- reacts with 1 mole of MES. We can say therefore:
(V_e mL) (0.100 M) base = (50.00 mL) (0.020 M) HA
V_e = 10.00 mL

The titration calculations for this problem can be handled through four steps:
- Before any base is added, the solution contains just HA in water. This is a weak acid problem in which the pH is determined by the equilibrium.

$$\text{HA} \quad \overset{K_a}{\rightleftharpoons} \quad \text{H}^+ \quad + \quad \text{A}^- \qquad\qquad (8.49)$$

- From the first addition of NaOH until immediately before equivalence point, there is a mixture of unreacted HA plus the A$^-$ produced by reaction in (8.48). We can use Henderson-Hasselbalch expression (8.44) to find pH.

- At equivalence point, all the HA has been converted into A$^-$. The problem is the same as if the solution had been made by just dissolving A$^-$ in water. We have a weak-base problem in which the pH is determined by the reaction (8.50).

$$\text{A}^- \; + \; \text{H}_2\text{O} \quad \overset{K_b}{\rightleftharpoons} \quad \text{HA} \; + \; \text{OH}^- \qquad\qquad (8.50)$$

- Beyond the equivalence point, excess NaOH is being added to a solution of A$^-$. We calculate pH as if we had simply added excess NaOH to water. We will ignore the very small effect from having A$^-$ present as well.

Region 1: Before Base is Added.
Before adding any base, we have a solution of 0.020 M HA with pKa = 6.27. This is simply a weak - acid problem.

$$Ka = \frac{x^2}{0.020 \; - \; x} = 1.0_3 \; x \; 10^{-4}$$

$$pH = -\log Ka = -\log 1.0_3 \; x \; 10^{-4} = 3.99$$

Region 2: Before the equivalence point
Upon adding OH$^-$, a mixture of HA and A$^-$ is created by the titration reaction (8.48). This mixture is a buffer whose pH can be calculated with the use of Henderson – Hasselbalch

$$pH = pKa + \log\left(\frac{[conjugate\ base]}{[acid]}\right)$$
(8.51)
Consider a point where 3.00 mL of OH$^-$ has been added

Once we know the quotient of [A⁻] / [HA] in any solution, we can easily calculate its pH using expression (8.44).

$$pH = 6.27 + log\left(\frac{0.300}{0.700}\right) = 5.90$$

It is obvious that in equation 8.51 we are using mmol because volumes cancel in the fraction [A⁻] / [HA]. The point at which the volume of titrant is $1/2\ V_e$ requires special attention in any titration.

$$pH = pKa + log\left(\frac{0.500}{0.500}\right)$$

$$pH = pKa \tag{8.52}$$

When the volume of titrant is $1/2\ V_e$, pH = pKa for acid HA. From the experimental titration curve below (Figure 8.2) you can find pKa by reading the pH when $V_b = 1/2\ V_e$ where V_b = volume of added base.

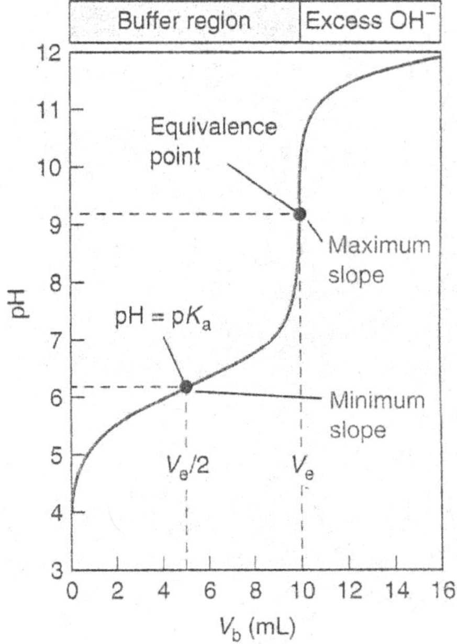

159

Once a mixture of HA and A- is precisely well recognized in any solution, this will mean the buffer is known. You can proceed to calculate the pH from the quoetient [A-] / [HA] expression (8.51).

Region 3: At the Equivalent Point
At the equivalence point (V_b = 10.00 mL), the quantity of NaOH is exactly enough to consume that HA.

The resulting solution contains 'only' A-; one can prepare the same solution by dissolving the salt Na^+A^- in water. It is apparent that Na^+A^- is a weak base. For the purpose of accomplishing the pH of weak base, it is important to write the reaction of the base with water:

$$A^- \quad + \quad H_2O \quad \rightleftharpoons \quad HA \quad + \quad OH^- \qquad\qquad K_b = K_w/K_a \qquad (8.53)$$

$$F' - x \qquad\qquad\qquad\qquad\qquad x \qquad\qquad x$$

The tricky point to be noted here is that the formal concentration of A- is no longer 0.020 M, which was the initial concentration of HA. The initial 1.00 mmol of HA in 50.00 mL has been diluted with 10.00 mL of titrant, thus:

$$[A^-] = \frac{1.00 \; mmol}{(50.00 \; + \; 10.00)mL} = 0.01667 \; M \; \equiv F'$$

Assume the formal concentration of A- is equal F', one can calculate pH from reaction equation (8.53).

$$K_b = \frac{K_w}{K_a} = \frac{x^2}{F' - x} = 1.8_6 x \; 10^{-8} \;\rightarrow\; x = 1.7_6 \times 10^{-5} \; M$$

$$pH = -\log [H^+]$$

$$pH = -\log \left(\frac{K_w}{x}\right) = 9.25$$

The pH at the equivalence point in this titration is 9.25 and not 7.00. The pH at equivalence point will often be above 7 for the titration of a weak acid with a strong base, because the acid is converted into its conjugated base at the equivalence point.

Region 4: After the Equivalence Point
At this juncture we are adding NaOH to a solution of A-. The base NaOH is so much stronger than the base A- that it is a fair approximation to say that the pH is determined by the concentration of excess OH- in the solution.

Let us calculate the pH when V_b = 10.10 mL, which is just 0.10 mL past V_e. The quantity of excess OH- is (0.100 mL) (0.100 M) = 0.010 mmol, and the total volume of solution is 50.00 + 10.10 mL = 60.10 mL

$$[OH^-] = \frac{0.010 \; mmol}{(50.00 + 10.10)mL} = 1.66 \times 10^{-4} \; M$$

$$pH = -\log\left(\frac{K_w}{OH^-}\right) = 10.22$$

8.14.3 Weak Diprotic Acid-Strong Base

A good example of diprotic acid is oxalic acid, $H_2C_2O_4$. This is a weak diprotic acid, which has the following ionization expression:

$$H_2C_2O_4 \text{ (aq)} \quad + \quad H_2O \text{ (l)} \quad \rightleftharpoons \quad H_3O^+ \text{ (aq)} \quad + \quad HC_2O_4^- \text{ (aq)} \tag{8.54}$$

$$K_{a1} = 5.9 \times 10^{-2}$$

$$HC_2O_4^- \text{ (aq)} \quad + \quad H_2O \text{ (l)} \quad \rightleftharpoons \quad H_3O^+ \text{ (aq)} \quad + \quad C_2O_4^{2-} \text{ (aq)} \tag{8.55}$$

$$K_{a2} = 6.4 \times 10^{-5}$$

The K_{a1} value is relatively large, which means there will be large increase in pH near the first equivalence point because there is buffering action on both sides of the first equivalence point. In such case there is buffering due to the existence of $H_2C_2O_4/HC_2O_4^-$ conjugate pair prior to the first equivalence point, and buffering caused by $HC_2O_4^-/C_2O_4^{2-}$ conjugate pair following the first equivalence point. This means that the species in solution tend to resist a rapid change in pH at the first equivalence point, and larger increase in pH at the second equivalence point is observed, despite the fact that $K_{a2} <$ K_{a1}. In fact, such increase in pH at the second equivalence point is due to hydrolysis of the $C_2O_4^{2-}$ ion that is formed following the first equivalence point.

$$C_2O_4^{2-} \text{ (aq)} \quad + \quad H_2O \text{ (l)} \quad \rightleftharpoons \quad OH^- \text{ (aq)} \quad + \quad HC_2O_4^- \text{ (aq)} \tag{8.56}$$

$$K_b = \frac{K_w}{K_{a2}} = 1.6 \times 10^{-10}$$

This justifies the reason for pH at the second equivalence point is > 7.

8.14.4 Titration of Weak Base with Strong Acid

The titration of a weak base with a strong acid is just the reverse of the titration of a weak acid with a strong base. Under the above title one should bear in mind the following:

- At the start point (0.00 mL acid added) pH is due to the [OH-] formed during ionization of the weak base.
- Any point between the starting point and the equivalence point is caused by the B/BH+ buffer that is created as the weak base, B, is often converted to its conjugate acid, BH+, during the reaction with the strong acid. The pH can be calculated by substituting the moles of base remaining and the mole of conjugate acid formed into expression (8.57).

$$[OH^-] = \frac{[base]}{[conjugate\ acid]} \cdot K_b \tag{8.57}$$

The above quantities can be calculated by using expressions 8.58 and 8.59.

mol base remaining = original mol base – mol acid added, and $\qquad$ (8.58)

mol conjugate acid formed = mol acid added $\qquad$ (8.59)

where mol base = mL $_{base}$. M_{base} and mol acid = mL$_{acid}$. M_{acid}.

- The pH at equivalence point is caused by hydrolysis of the conjugate weak acid cation, BH$^+$, which is formed during titration.

$$BH^+ (aq) \quad + \quad H_2O (l) \quad \rightleftharpoons \quad H_3O^+ (aq) \quad + \quad B (aq) \tag{8.60}$$

- In fact, the pH beyond the equivalence point is due to [H$_3$O$^+$] formed by 100% ionization of the excess strong acid.

$$[H_3O^+] = \frac{moles\ of\ acid\ added - original\ moles\ of\ base}{volume\ of\ base(mL) + \ volume\ of\ acid\ added} \tag{8.61}$$

Because the reactants are weak acid versus strong base, the reaction goes essentially to completion after each addition of acid. There are four distinct regions of the titration curve:

- Before acid is added, the solution contains just the weak base, B, in water. The pH is determined by the K_b in water. The pH is determined by the K_b reaction:

$$B \quad + \quad H_2O \quad \overset{K_b}{\rightleftharpoons} \quad BH^+ \quad + \quad OH^-$$

$F - x \qquad\qquad\qquad\qquad\qquad x \qquad\qquad x$

- Between the initial point and the equivalence point, there is a mixture of B and BH$^+$ (a buffer),

The pH is computed by using the expression hereunder:

$$pH = pKa\ (for\ BH^+) + \log\left(\frac{[B]}{BH^+}\right)$$

At the special point where $V_a = V_e / 2$, pH = pKa (for BH$^+$)

- At the equivalence point, B has been converted into BH$^+$, a weak acid. The pH is calculated by considering the acid dissociation reaction of BH$^+$:

$$BH^+ \quad \rightleftharpoons \quad B \quad + \quad H^+ \qquad Ka = K_w/K_b$$

$F' - x \qquad\qquad\qquad x \qquad\qquad x$

The formal concentration of BH$^+$, F', is not the same as the original formal concentration of B, because there has been some dilution. Because the solution contains BH$^+$ at the equivalent point, it is acidic. The pH at the equivalence point must be below 7.

- After the equivalence point, there is excess strong acid in the solution. This problem can be treated by just considering only the concentration of excess H$^+$ and ignoring the contribution of weak acid BH$^+$.

Example 8.3

Titration of Pyridine

Consider the titration of 25. 00 mL of 0.08364 M Pyridine with 0.1067 M HCl

Titration reaction

$$\text{Pyridine (B)} \quad + \quad H^+ \quad \longrightarrow \quad BH^+$$

$(V_e \text{ mL}) (0.1067 \text{ M}) = (25.00 \text{ mL}) (0.08364 \text{ M})$

$V_e = 19.60 \text{ mL}$

(a) Find the pH when V_a = 4.63 mL, which is before the equivalence point, and

(b) Find the pH at the equivalence point.

Solution:

(a) At 4.63 mL part of the pyridine has been neutralized, so there is a mixture of Pyridine and Pyridinium ion, a buffer. The initial millimoles of pyridine are (25.00 mL) (0.08364 M) = 2.09 mmol.

(b) The added H$^+$ is (4.63 mL) (0.1067 M) = 0.494 mmol. Therefore, we can solve this problem as follows:

Titration reaction	B	+	H$^+$	$\longrightarrow$	BH$^+$
Initial mmol	2.091		0.494		-
Final mmol	1.597		-		0.494

$$pH = pK_{BH^+} + log\left(\frac{[B]}{[BH^+]}\right) = 5.20 + log\left(\frac{1.597}{0.494}\right)$$

$$pH = 5.71$$

At the equivalence point (19.60 mL) enough acid has been added to convert all the pyridine (B) into BH$^+$. The pH is governed by dissociation of the weak acid. BH$^+$, whose acid dissociation constant is given in the expression hereunder.

$$K_a = \frac{K_w}{K_b} = 6.3 \; x \; 10^{-6}$$

The formal concentration of BH^+ is equal to the initial mmol of Pyridine divided   by mL of solution at the equivalence point.

$$F = \frac{(2.09) \; mmol}{(25.00 + 19.60)mL} = 0.04688 \; M$$

$$BH^+ \rightleftharpoons \qquad B \quad + \quad H^+ \qquad\qquad Ka = 6.3 \; x \; 10^{-6}$$

$$F' - x \qquad\qquad\qquad x \qquad\qquad x$$

$$\frac{x^2}{F-x} = \frac{x^2}{0.04688 - x} = K_a = 6.3 \; x \; 10^{-6} \;\; \rightarrow \;\; x = [H^+] = 5.4 \; x \; 10^{-4} \; M$$

$$pH = -\log[H^+] = 3.27$$

The pH at equivalence point is acidic because the weak base has been converted into a weak acid.

Practice 8.2

What is the equivalence volume in the titration of 100.00 mL of 0.10 M cocaine ($K_b = 2.6 \; x \; 10^{-6}$) with 0.20 M HNO_3? Calculate the pH at $Va = 0.0$, and 20 mL.

Solution:

pH at $Va = 0.0$ mL is 10.71 and at $Va = 20.0$ mL is 4.80.

8.15   Acid-Base Indicators

An indicator is a compound that can change form in response to the titrant concentration which results in two forms of the compound with different colours. Indicators and pH measurements are commonly used to find the end point in an acid-base titration. The intention of an acid-base titration is to add an amount of acid or base that is stoichiometrically equivalent to the amount of base or acid being titrated. This happens at the equivalence point of the reaction. The titration curve is used to determine equivalence point when reached, because the equivalence point corresponds to the midpoint of the rapid pH raise. In actual fact, acid-base indicators are commonly used to estimate precisely the equivalent point. An acid-base indicator is the substance that often exists in conjugate acid and base forms, having different colours and has the characteristic of changing colour in a given pH range. We say the colour change occurs at the end point of the titration, because that is the point at which we end the titration, and we carefully select an indicator that will keep the volume difference between the end point and equivalence point negligibly small.

Commonly acid-base indicator is organic dye that is a weak acid and can be represented by a general form of Hind, as in expression 8.62 and 8.63 or 8.64

$$HInd \text{ (aq)} \quad + \quad H_2O \text{ (l)} \quad \rightleftharpoons \quad H_3O^+ \text{ (aq)} \quad + \quad Ind^- \text{ (aq)} \tag{8.62}$$

and

$$K_a\, HInd = \frac{[HInd]\,[H_3O^+]}{[Ind^-]} \tag{8.63}$$

$$[H_3O^+] = \frac{[HInd]}{[Ind^-]} \cdot K_a\, HInd \tag{8.64}$$

which tells us the ratio hereunder,

$\dfrac{[HInd]}{[Ind^-]}$ is controlled by the $[H_3O^+]$ and pH of the solution. The latter ratio is important because it is useful in the determination of the solution colour. When the ratio $[HInd]\,/\,[Ind^-] = 10\,/\,1$, the colour of Ind^- predominates, and when $[HInd]\,/\,[Ind^-] = 1\,/\,10$, the colour of Ind^- predominates. This tells us that the colour change occurs over a hydronium ion concentration which is about 100; this corresponds to 2 pH units, and the pH range of an indicator is given by expression 8.65:

$$pH\ range\ of\ HInd = pK_a\, HInd \pm 1 \tag{8.65}$$

It is important to remember that there is a variety of indicators available with differing pH ranges. For example, phenolphthalein, which changes from colourless to pink in the pH range of 8 to 10, is suitable for use with strong acid-strong base, strong base-strong acid and weak acid-strong base titrations, due to the fact that the pH at the equivalence point for the first two types is 7, and that for the latter pH is > 7 due to anion hydrolysis. Nevertheless, methyl red, which changes in the pH range 4.2 to 6.2, is suitable for use with a weak base-strong acid titration, because the pH at equivalence point is < 7 due to cation hydrolysis.

In complexometric titrations, for example, a commonly used indicator is a ligand that has different colours depending of whether it is complexed with coordination center. For such a case, a good example commonly used is Eriochrome Black T (EBT). This type of indicator, as can be seen in chapter 6, section 6.2.4, is a weak acid. The anion part of sodium salt has two protons to donate. The following expressions (8.66 and 8.67) show the reaction with water and their respective equilibrium constants.

$$H_2In^- \quad + \quad H_2O \;\rightleftharpoons\; HIn^{2-} \quad + \quad H_3O^+ \qquad K'_{a2} = 5\,x\,10^{-7} \tag{8.66}$$

$$HIn^{2-} \quad + \quad H_2O \;\rightleftharpoons\; HIn^{3-} \quad + \quad H_3O^+ \qquad K'_{a3} = 2.8\,x\,10^{-12} \tag{8.67}$$

The above three conjugate acids of the indicator have different colours. Starting with the most acidic (H_2In^-), which is red, followed by (HIn^{2-}) which is blue, and the most basic (In^{3-}) is orange. This observation means that EBT can conveniently also be used in acid-base titration indicator. For most of metals, the colour of the EBT is red. It is important to note that the colour of the indicator depends on the metal ion concentration. Suppose the metal is the analyte and the complexing agent is titrant, the indicator is in its metal complex form red before the end point due to the excess of the metal ion, and when an excess of complexing agent has been added, in its uncomplexed form blue at pH 10. If it happens pH is below 6.0 uncomplexed indicator is in acidic form, which has the same colour as the complex.

Suppose you are using Eriochrom Black T in the titration of calcium with EDTA. The Ca^{2+}-EBT complex formation constant is 2.5×10^5, and the actual titration is carried out at pH 10.00. At that particular pH, the uncomplexed EBT is almost completely in the blue HIn^{2-} form. The red complex has $CaIn^-$ form. Therefore, the overall reaction with calcium is the combination of both deprotonated reaction of HIn^{2-} and the complexation reaction In^{3-} in expression 8.68:

$$\left. \begin{array}{l} H_2In^- \;+\; H_2O \;\rightleftharpoons\; HIn^{2-} \;+\; H_3O^+ \\[2em] Ca^{2+} \;+\; In^{3-} \;\rightleftharpoons\; CaIn^- \\[0.5em] \hline \\ Ca^{2+} \;+\; HIn^{2-} \;+\; H_2O \;\rightleftharpoons\; CaIn^- \;+\; H_3O^+ \\ \quad\quad\quad\quad \text{blue} \quad\quad\quad\quad\quad\quad \text{red} \end{array} \right\} \tag{8.68}$$

for which the formal equilibrium constant can be expressed as in expression 8.69.

$$K'_{eq} = \frac{[CaIn^-]\,[H_3O^+]}{[Ca^{2+}]\,[HIn^{2-}]} = K'_{a3}\,K'_{CaIn} \tag{8.69}$$

In a situation where red and blue colours are of equal intensity, this will result in colour changes when the concentration of $CaIn^-$ and HIn^{2-} are equal. Upon substituting this condition to expression 8.69, the following expression 8.70 is obtained:

$$[Ca^{2+}] = \frac{[H_3O^+]}{K'_{a3}\,K'_{CaIn}} \tag{8.70}$$

It is interesting to note that formation constant of Ca^{2+} with EDTA is greater than that of Mg^{2+} with EDTA, because of the fact that the EDTA reacts with the Ca^{2+} first, and once this appears to be almost gone, the EDTA at last reacts with the Mg^{2+}. From the fact that EBT is a suitable indicator for the Mg^{2+} titration, it can be used in titration of the mixtures of Ca^{2+} and Mg^{2+}, though it is not suitable for the titration of Ca^{2+} when it is alone.

Suppose the mixture of Ca^{2+} and Mg^{2+} is titrated with EDTA using the Erichrome Black T indicator, the equivalent point is attained when the amount of EDTA added is equal to the sum of the amounts of Ca^{2+} and Mg^{2+}. Obviously, individual amounts are not determined in this way. In the assessment of total water hardness, the sum total of Ca^{2+} and Mg^{2+} are useful.

8.16 Problems

8.16.1. The concentration of OH^- in a sample of seawater is 5.0×10^{-6} M. Calculate the concentration of H_3O^+ ions, and classify the solution as acidic, basic or neutral.

8.16.2. At 50 °C the value of K_w is 5.5×10^{-14}. What are the concentrations of H_3O^+ and OH^- in a neutral solution at 50 °C?

8.16.3. (a) Why is the equivalence-point pH necessary below 7 when a weak base is titrated with strong acid?
(b) Consider the titration of 25.0 mL of 0.05 M $HClO_4$ with 0.10 M KOH. Find the equivalence volume. Find the pH at the following volume 0.00 and 1.00 mL.

8.16.4. Consider the titration of 100.00 mL of 0.01 M NaOH with 1.00 M HBr. What is the equivalence volume? Find the pH at volumes of Va = 0.0 and 5.00 mL.

8.16.5. What is the pH at the equivalence point when 0.10 M hydroxyacetic acid is titrated with 0.05 M KOH?

8.16.6. When 16.24 mL of 0.064 M KOH was added to 25.00 mL of 0.094 M weak acid, HA, the observed pH was 3.62. Find pKa for the acid.

8.16.7. From chemistry point of view it is not safe to mix acid with cyanide (CN^-), because it produces HCN (g). But, just for fun, calculate the pH of a solution made by mixing 50.00 mL of 0.10 M NaCN with 4.20 mL of 0.44 M $HClO_4$.

8.16.8. A 50.0 mL volume of 0.05 weak acid HA (pKa = 4.00) was titrated with 0.50 M NaOH. Write the titration reaction and find V_e. Find the pH at V_b = 0.0 and 4.00 mL.

8.16.9. Provide an explanation for what chemistry occurs in each region of the titration of weak acid, HA, with OH^-. State how you would calculate the pH in each region.

8.16.10. Phenolphthalein is used as an indicator for the titration of HCl with NaOH.
(a) What colour change is observed at the end point?
(b) The basic solution just after the end point slowly absorbs CO_2 from the air and becomes more acidic by virtue of the reactionhereunder;

$$CO_2 \quad + \quad OH^- \quad \rightleftharpoons \quad HCO_3^-$$

This causes the colour to fade from pink to colourless. If you carry out the titration too slowly, does this reaction lead to a systematic or a random error in finding the end point?

8.16.11. A solution of 100.00 mL of 0.04 M sodium propanoate (the sodium salt of propanoic acid) was titrated with 0.084 HCl. Find V_e and calculate the pH at $Va = 0$, and $1/4V_e$.

8.16.12. Consider the titration of 100 mL of 0.016 M HOCl ($Ka = 3.5 \times 10^{-8}$) with 0.040 M NaOH. How many milliliters of 0.040 M NaOH are required to reach the equivalence point? Calculate the pH:
(a) After addition of 10.0 mL of 0.040 M NaOH.
(b) Half way to the equivalence point.
(c) At the equivalence point.

8.16.13. (a) What are the qualities of indicator for a complexation titration?
(b) How does the pH of titration solution affect the choice(s) or requirements of the indicator?

References

Brent, L. (2009). *Acids and bases*, New York, NY: Crabtree Publisher.

Clareen, S. S., Marshall, S. S. R., Price, K. E., Royall, M. B., Yoder, C. H. and Schaeffer, R. W. (2000). "The Synthesis and Analysis of Ammine Complexes of Copper and Silver Sulfate," *J. Chem. Ed. 77*, 904.

Clayden, J., Warren, S. *et al.*, (2000). *Organic Chemistry* Oxford University Press.

Crossno, S. K., Kalbus, L. H. and Kalbus, G. E. (1996). "Determinations of Carbon Dioxide by Titration" *J. Chem. Ed. 73*, 175.

Davis, C. M. and Mauck, M. C. (2003). "Titrimetric Determination of Carbon Dioxide in a Heterogeneous Sample ('Pop Rocks')," *J. Chem. Ed. 80*, 552.

Davranches, M., Lacour, S., Boardas, F. and Bollinger, J. C. (2003). "An Easy Determination of the Surface Chemical Properties of Simple and Natural Solids" *J. Chem. Ed. 80*, 76.

Eugene, L. (2002). Chemistry, Upper Saddle River, New Jersey Prentice-Hall.

Lake, M. E., Grunow, D. A. and Su, M. C. (1992). Graphical Presentation of Acid-Base Re actions Using a Computer-Interface Autotitration," *J. Chem. Ed. 69*, 299.

MacMurry, J. E. and Fay, R. C. (2008). *Chemistry* 5th, Upper Saddle River, NJ.

Meyers, R. (2003). *The Basic Chemistry*. Greenwood Press.

Oxlade, C. (2002). *Acids and bases*, Chicago, IL: Heinemann Library.

Petrucci, H. R., William, S. H., Herrring, F. G. and Jeffrey, M. D. (2007). *General Chemistry Principles and Modern Applications*. 9th Upper Saddle River, New Jersey Pearson Prentice Hall.

Tucker, S. A. and Acree, Jr., W. E. (1994). "A student-Designed Analytical Laboratory Method: Titration and Indicator Ranges in Mixed Aqueous-Organic Solvents," *J. Chem. Ed. 71*, 71.

Williams, K. R and Tennant, L. H. (2001). "Micelles in Physical/Analytical Chemistry Laboratory: Acid Dissociation of Neutral Red Indicator," *J. Chem. Ed. 78*, 349.

Chapter 9

PRINCIPLES OF CHROMATOGRAPHY

9.1   Introduction

Chromatography is a separation process which is based on the flow of a fluid sample through a column, whereby finely divided solids serve as a support for liquid phase. Generally speaking, different solutes show variations in the way they partition between the stationary solid phase and the mobile fluid phase (liquid solution or gas), which results in bands being formed at different positions in the column. Separation of the bands requires careful selection of the fluid to be added in the column after the sample has been added: this process is known as *elution*. In Figure 9.1 a mixture of compounds A and B is added to the column, which is packed with solid particles and filled with appropriate solvent. Upon opening the outlet, A and B flow down the column based on their respective affinities with stationary phase. Both A and B will be washed accordingly with fresh solvent applied on the top of the column.

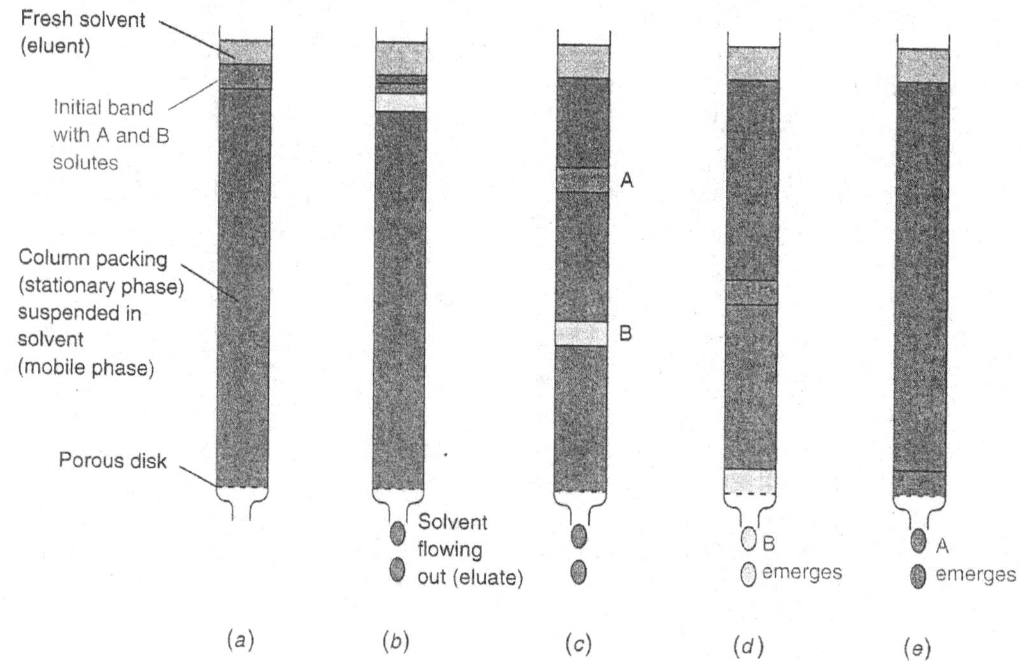

Figure 9.1: Separation of Solute A and B Based on Their Respective Affinities

For the case of the above example solute A is more strongly adsorbed to the solid particle than solute B; this means that solute A will spend much more time in the column than solute B, which will emerge at the bottom first, as can be seen in Figure 9.1 part (d). The following are some important terms which are commonly used in this area:

Mobile phase (solvent moving through the column) in chromatography is either a solid or gas.

The *stationary phase* (substance that stays fixed in the column) can be solid or liquid which often is covalently bonded to solid particles or to the inside of the wall of hollow capillary column. It should be clearly noted that in gas chromatography the mobile phase is a gas, and in liquid chromatography, the mobile phase is liquid.

Eluent this is the fluid entering the column.

Eluate this is the fluid exiting the column, and

Elution this is the process of passing liquid or gas through a chromatography column.

9.2   Types of Chromatography

Hereunder are the types of chromatography with variation in the way the solute is interacting with the stationary phase:

(i) Adsorption Chromatography

This is a type of chromatography that uses a solid stationary phase, and a liquid or gaseous mobile phase. The solute is absorbed on the surface of the solid particles as in Figure 9.2. This method is commonly useful in preparative and analytical organic chemistry and biochemistry. Under the conditions that all factors are constant, the adsorbability of organic compound increases with its polar character; this means the number and kind of polar groups in the molecule matters. Furthermore, the adsorbability often increases with an increase in molecular weight and molecular volume. The stronger the polar solvent, the less strongly the solute is adsorbed, so that the order of decreasing adsorption with solvent is as follows: petroleum ether, carbon tetrachloride, ether (anhydrous and free from alcohol), benzene, chloroform, alcohols and water.

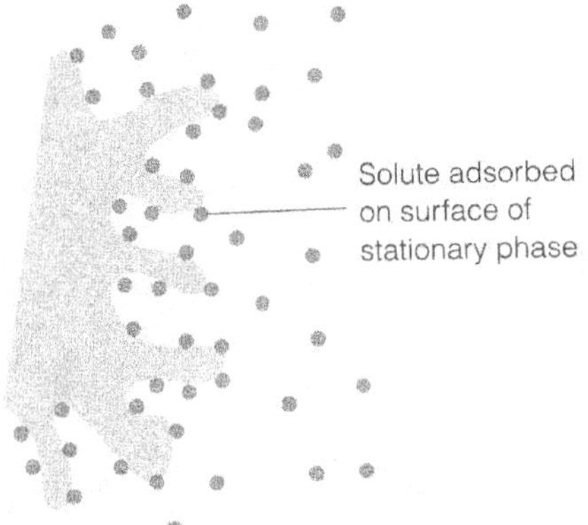

Figure 9.2: Adsorption Chromatography

(ii) Partition Chromatography

This basically involves a thin liquid stationary phase coated on the surface of a solid support. This set up solute equilibration between the stationary liquid and mobile phase: see Figure 9.3. This type of chromatography is based on the difference in distribution of different solutes between two immiscible liquid phases. Most of inorganic partition chromatographic separations are carried out with cellulose (pulp or powder) as the column filler. For example, uranium and thorium can be separated from each other and from most other elements by using a cellulose column and an ether solution containing nitric acid as the mobile phase. Uranium is always eluted with 97 percent ether-3 per cent concentrated nitric acid and thorium with ether containing 13 percent nitric acid. Another interesting aspect is in uranyl and thorium nitrates being more soluble in the ether phase than in the aqueous phase, and this makes it move fast down the column, whereas most of the other metal nitrates move relatively slowly. In fact, trace metals like lead, zinc, manganese, and other heavy metals reacting with dithizone (Figure 9.4) can be isolated from natural waters by running the sample through a column of cellulose acetate holding a carbon tetrachloride solution of dithizone as a stationary state. Under this condition, metals in the column can be eluted by using acid and ammonia.

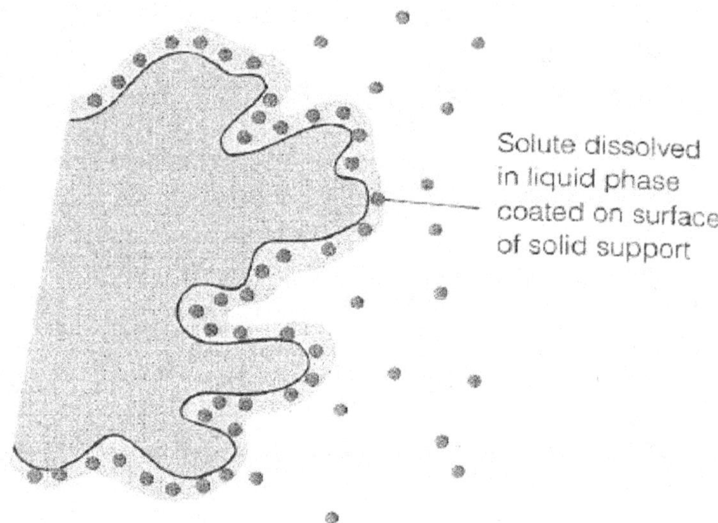

Figure 9.3: Partition Chromatography

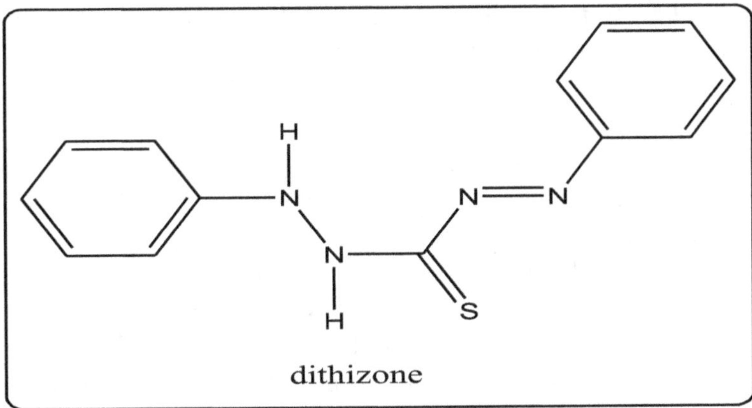

Figure 9.4: Structure for Dithizone

(iii) Molecular Exclusion Chromatography

This separates molecules according to their respective size, with larger molecules passing through the column most quickly, the reason being that there is no attraction between solute and stationary phase. The stationary phase has pores small enough to exclude large molecules, but not small ones: see Figure 9.5. This can be closely and clearly observed in column (a) and (b).

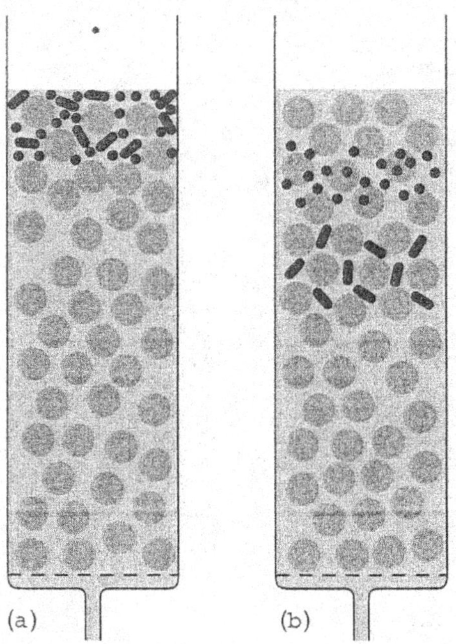

Figure 9.5: Molecular Exclusion Chromatography

As you can see, in column (b) large molecules stream past without entering the pores. Small molecules as can be seen in the same column are entering

173

the pores and thus must flow through large volume before they get out of the column. That is why they will be collected as last collection or extract.

(iv) Affinity Chromatography

This technique is considered to be the most selective type of chromatography which is used to isolate a single compound from a complex mixture. The technique is based on the specific binding of a given compound to the stationary phase: see Figure 9.6. When the sample is allowed to pass through the column, selectively one solute is bound. Once everything else has been washed through the column, the adhered solute is eluted by adjusting conditions such as pH or ionic strength to weaken its binding. This separation technique is most applicable in biochemistry and is based on specific interactions between enzymes and substrate, antibodies and antigens or receptors and hormones. A good example of how this works is in protein immunoglobulin G (IgG), which can be separated by affinity chromatography on a column containing covalently bound protein Y. The protein Y binds in the specific region of IgG in the column at pH $\geq$ 7.2. Upon passing crude mixture containing IgG and other proteins at pH 7.6 in the column, everything except IgG was eluted within eighteen (18) seconds. After sixty seconds the eluent pH was lowered to 2.6 and IgG was carefully eluted in seventy eight (78) seconds.

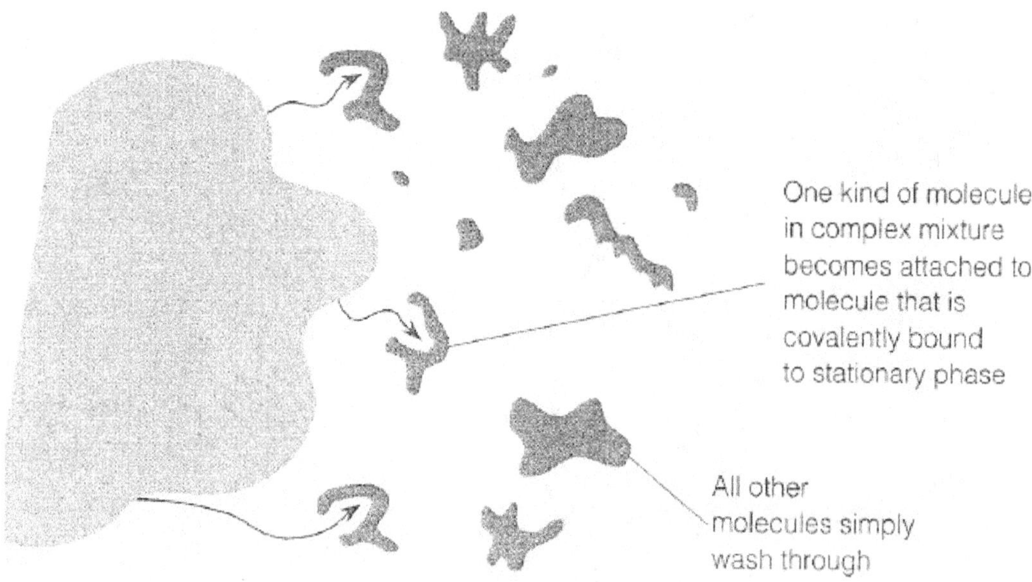

Figure 9.6: Affinity Chromatography

(v) Gas Chromatography

This technique is divided into gas-solid chromatography and gas-liquid chromatography, and the latter is relatively more important. For the case of gas-liquid chromatography, the stationary phase is a liquid held on a solid support such as diatomaceous earth, the gas or vapour partitions between the non-volatile liquid phase and the inert carrier gas (for example, He_2, N_2 or H_2). Inside gas chromatography, the gaseous components form a series of bands in the column, which are eluted by the stream of carrier gas and can be detected as well as determined in the stream of carrier gas by their effect on thermal conductivity, ionization, as well as other known gas properties. The mobile gaseous phase transports gaseous solute through a long, thin column containing stationary phase as in Figure 9.7. The process in Figure 9.7 begins by injecting a volatile liquid through an injector port into a heated port, which vapourizes the sample. The injected sample is swept through the column by carrier gas (He_2, N_2 or H_2), and the separated solutes flow through a detector, whose response is detected and recorded by the computer. For that to happen, the column should be hot enough to produce reasonable vapour pressure for each solute to be eluted in appropriate time.

The detector often is kept at a higher temperature than that of the column so that all solutes are in gaseous form. Typically, the amount of sample to be injected ranges between 0.1-2 µL for analytical chromatography, and preparative columns can handle between 20-1000 µL. Indeed, the gaseous compounds are introduced in volumes of 0.5-10 mL by gas tight syringe or a gas-sampling valve.

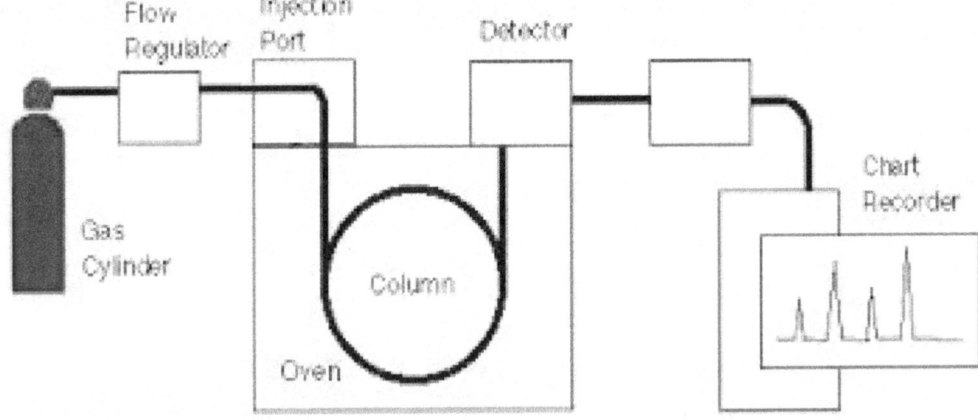

Figure 9.7: Schematic Representation of a Gas Chromatography

(vi) Classical Liquid Chromatography

In Figure 9.1 shows an example of modern chromatography, in which a sample is added to the top of an open, and gravity-feeds the column which is containing stationary phase. This follows the descriptions of high performance liquid chromatography, which uses a well-closed column under high pressure and is currently the common form of chromatography. Nevertheless, open columns are used for preparative separations in both biochemistry and chemical analysis.

Usually, the action of adding sample uniformly on the column is an art; therefore, applying sample evenly, and obtaining symmetric elution bands requires special attention. For the case of stationary solid phase the sample is often poured into the column after preparation of *slurry* (a mixture of solid and liquid) and the slurry is carefully added down the wall of the column. When you are doing the latter, it is important to ensure that there is no formation of distinct layers, which often form when some of the *slurry* is allowed to settle before more is poured in.

The solvent below the top of the stationary phase should not be drained, as this will avoid air spaces and irregular flow patterns will be created. The solvent should be meticulously directed down the wall of the column. At all times solvents should not be allowed to dig a channel into a stationary phase. It should be clearly noted that a slow flow rate is a prerequisite for maximum resolution.

Solvents

It is a normal tendency for solvent to compete with solute for adsorption sites on the stationary phase of adsorption chromatography. It has been established that the abilities of different solvents to elute certain solute from the column are almost independent of the nature of the solute. Elution is displacement of solute from the adsorbent by solvent as is described in Figure 9.8.

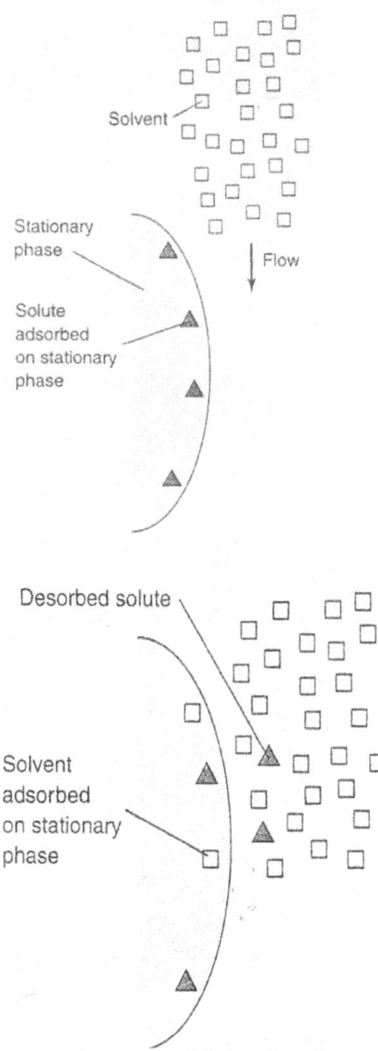

Figure 9.8: Solvent Molecules Compete with Solute Molecules for Binding Sites

The ranking of solvents by their relative abilities to displace solutes from a given adsorbent is called *eluotropic series*. Table 9.1 shows eluent strength, which measures solvent adsorption energy, whereby the value of pentane has been given the value of zero (0). The more polar the solvent, the greater is eluent strength. The greater the eluent strength, the more rapidly the solute will be eluted from the column. The eluent strength gradient is mostly used for many separations. This includes weakly retained solutes which are eluted with a solvent of low eluent strength. Also, when two solvents are mixed, either in discrete steps or continuously, they improve eluent strength and elute more strongly adsorbed solute. Just a small amount of polar solvent significantly increases the eluent strength of a nonpolar solvent.

Table 9.1: Eluotropic Series and Ultraviolet Cutoff Wavelength of Solvents for Adsorption Chromatography on Silica

Solvent	Eluent strength (ε^o)	Ultraviolet cutoff (mm)
Pentane	0.00	190
Hexane	0.01	195
Heptaine	0.01	200
Trichlorotrifluoroethane	0.02	231
Toluene	0.22	284
Chloroform	0.26	245
Dichloromethane	0.30	233
Diethyl ether	0.43	215
Ethyl acetate	0.48	256
Methyl *t*-butyl ether	0.48	210
Dioxane	0.51	215
2-Propanol	0.60	205
Methanol	0.70	205

Source: From: Snyder, L. R. (1983). High-Performance Liquid Chromatography, Academic Press New York, Burdick and Jackson, (1990). Solvent Guide, 3rd ed. Burdick and Jackson Laboratories.

(vii) High-Performance Liquid Chromatography (HPLC)

This operates under high pressure to force eluent through a closed column packed with micrometer-size particles that provide intense separations. HPLC equipment uses columns with diameters ranging between 1-5 mm and length of 5-30 cm, yielding 50,000 to 100,000 plates per meter, which is detailed in section 9.4.1. Some important components include a solvent delivery system, a sample injection valve, a detector, and a computer to display the results. This is one of the greatest techniques for separation, and at preparative industrial level it can handle up to 1 kg of sample at once.

In Figure 9.9 it can be observed how resolution has been improved with stationary phase particle size decrease as you move from (a) to (b). One can also note that the peaks are getting sharper as you move from (a) to (b) and how a decomposition product is resolved from the slow-moving component. Using fine particles usually has the drawback of being resistant to solvent flow. It is important to remember that the smaller particles of stationary phase improve the resolution by allowing solute to diffuse faster, and thereby speeding the equilibration to be reached faster, between the stationary and mobile phases.

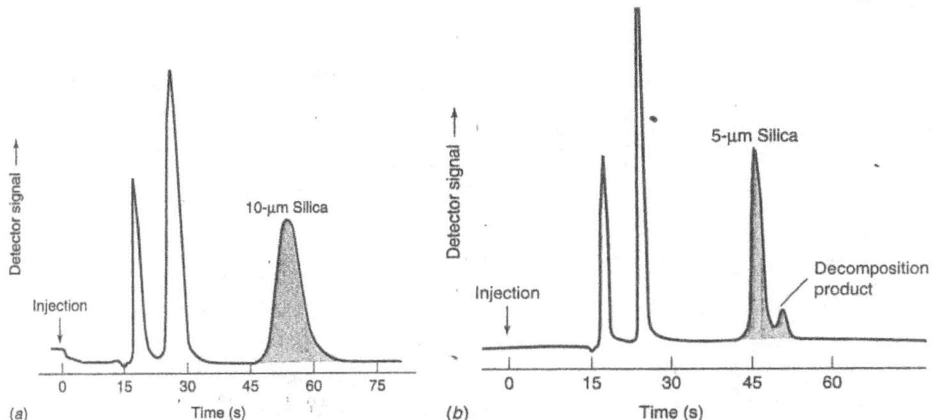

Figure 9.9: Chromatogram of the Same Sample Run on Columns with Different Particle Size
Source: From Majors, R. M. (1973). Effect of Particle Size on Column Efficiency in Liquid-Solid Chromatography, *J. Chromatogr. Sci. 11,* 88.

9.3   Stationary Phase

It should be noted that *normal-phase chromatography* uses a polar stationary phase and relatively less polar solvent. Typically, *reversed-phase chromatography* is a very common procedure, whereby stationary phase is non-polar or weakly polar and the solvent is more polar. Eluent strength can be improved by adding more polar solvent, as we have seen in Table 9.1. The advantage of *reverse phase* chromatography it removes tailing arising from adsorption of polar compounds by polar pickings: see Figure 9.10. In addition, reverse-phase chromatography has the drawback of being insensitive to polar impurities, for example water in the eluent.

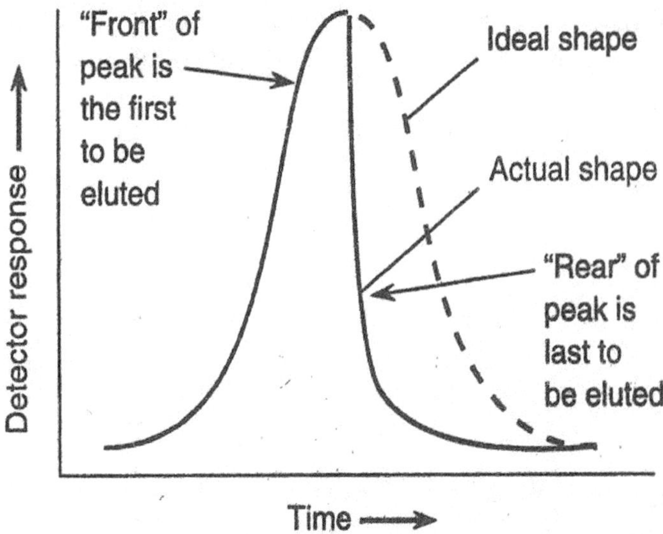

Figure 9.10: The Peak Shape with Normal Front and a Grossly Elongated Rear

The most commonly stationary phase support used are *microporous particles* of silica with diameter ranging between 1.5-10 µm. Such particles are permeable enough for solvent and have a surface area of about 500 m² per gram of silica. The adsorption of solute occurs directly on the silica surface. The octadecyl (C_{18}) stationary phase is, by far, the most commonly used in HPLC. The attachment of Si-O-Si bond through silica is stable only over the pH ranging between 2 to 8. It should be clearly noted that strongly acidic or basic eluents cannot be used with silica.

9.4   Sample Preparation for Chromatography

This preparation involves transformation of a sample into a suitable form for analysis. The process may entail extracting analyte from a complex matrix, preconcentrating very dilute analytes to get a high enough concentration to measure, removing or masking interfering species, or chemically transforming (derivaterizing) the analyte into a form that is easier to separate or to detect. All *solid microextraction, purge*, and *trap* are commonly used in preparation for gas chromatography samples. On the other hand, solid phase extraction is known to be the simplest technique for liquid chromatography sample preparation.

(i) Solid-phase Microextration

This is a simple method used to extract compounds to be used for gas chromatography from liquids, air, or even sludge without using any solvent. A fused silica fiber is coated with a 100 µm thick film of nonvolatile liquid stationary phase. Just a fraction of the analyte in the sample is extracted into the fiber. In fact, solid-phase microextration can be adapted to liquid chromatography by inserting the fiber into an injection port in which the fiber should be washed with a strong solvent see Figure 9.11.

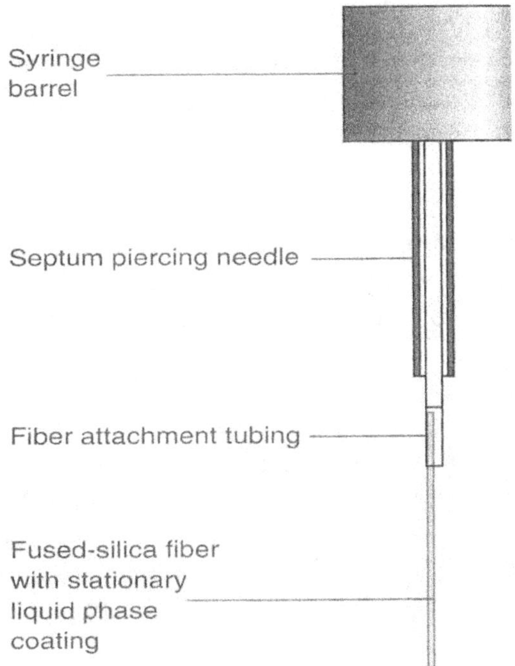

Figure 9. 11: Syringe for Solid-phase Microextraction

(ii) Purge and trap

This approach is used to remove volatile nalytes from liquid or solids (a good example being ground water or soil), concentrating the analyte and introducing it into chromatography. In comparison with solid-phase microextration, which removes just a fraction of analyte from the sample, the major purpose in purge and trap is total removal by 100% of the analyte from the sample. It is always difficult to remove quantitatively polar analytes from polar matrices.

(iii) Solid-phase extraction

This involves the use of liquid chromatography solid phase in a short open column to collect one or more analytes from a liquid mixture. A good example is when a solution containing nonpolar and weakly polar organic molecules is passed through solution C_{18}-silica: often the organic compounds are retained on the silica.

Also, when steroid is tested in the drug testing of athletes, urine is allowed to pass through a short C18-column and illegal steroids will be retained in the column. When urine constituents (polar and ionic substances) are allowed to pass through the column they will not be retained in the column. Upon washing the column with a small volume of solvents (hexane or dichloromethane), steroids are eluted and can be analyzed by chromatography.

(iv) Sample cleanup

This generally means the removal of unwanted components of an unknown that interferes with the measurement of analyte. In the analysis of urine-steroid, for example, polar molecules that are not captured by C_{18}-silica are washed through the column and are separated from the steroids.

Practice 9.1

(a) Why is it necessary to use cold trapping on the gas chromatography column if a sample is introduced from a solid-phase microextraction syringe or from a purge and trap absorption tube? (b) What is the purpose of solid-phase extraction?

9.5 Descriptions of Gas Chromatography

This shows the chromatography column response to a given solute as it exits. A *chromatogram* basically shows detector response as a function of time (or elution volume) in a chromatography experiment (Figure 9.12). As it can be seen in Figure 9.12, every peak represents a specific substance eluted from the column. The *retention time*, t_r, is the time needed for injected solute to reach the detector.

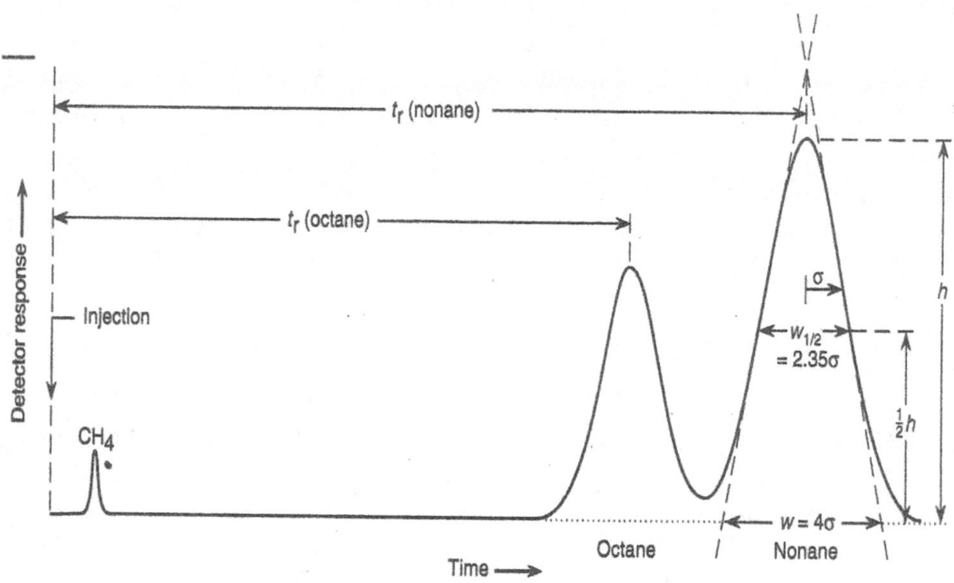

Figure 9.12: Schematic gas Chromatogram Showing t_r and $w_{1/2}$

9.5.1 Theoretical Plates

Hypothetical gas chromatographic peak has a Gaussian shape, like that in Figure 9.12. Suppose the height of the peak is h, the width at half-height is $w_{1/2}$, and is measured at ½ h. At half-height of Gaussian peak, $w_{1/2}$ is equal to

2.35σ, where σ stands for standard deviation of the peak. The width of the peak at baseline is equal to 4σ.

Many years ago, volatile compounds were separated by distillation process, which was considered to be a powerful method. A distillation column was divided into plates in which both liquid and vapour establish equilibrium with each other. It was noted that the more plates on a given column, the more equilibration steps, and the better separation between compounds with different boiling points. From distillation, the term theoretical plate has emerged and has become useful in chromatography. Consider a chromatography column as if it is divided into discrete sections called *theoretical plates* in which a solute molecule equilibrates between the mobile phase and stationary phase. The number of equilibration steps between injection and elution determines how long the compound can be retained in the column. The more theoretical plates, the narrower the bandwidth when the compound exits. The number of theoretical plates of each compound in Figure 9.12 can be calculated by measuring both retention time and the width at half- height:

$$N = \frac{5.55 \, t_r^2}{w_{1/2}^2}$$

$$(9.1)$$

Both retention time and width can be measured in the units of time or volume (such as millilitrers of eluate). In addition, for convenience, both t_r and $w_{1/2}$ in expression 9.1 must be expressed with the same units.

Suppose that in our mind we divide column into N theoretical plates, then the *plate height, H,* is the length of one plate. The value of H can be calculated by dividing the length of the column (L) by the number of theoretical plates on the column. Plate height therefore can be expressed as:

$$H = \frac{L}{N}$$

$$(9.2)$$

The above expression has revealed that the smaller the plate height, the narrower the peaks. As the plate height decreases, the ability of the column to separate the components of mixture is always improved. Typically, different solutes behave as if the column possesses different plate heights, because each solute equilibrates between the mobile and stationary phase at different rates. Commonly, plate heights range from 0.1 to 1 mm in gas chromatography, 10 μm in high performance chromatography, and < 1 μm in capillary electrophoresis (see section 9.4).

Example 9.1
A solute with a retention time of 360 s has a width at half-height of 6 s on a column 10 m long. Calculate the number of plates and plate height.

Solution:

$$N = \frac{5.55\, t_r^2}{w_{1/2}^2}$$

$$N = \frac{5.55 \times 360^2}{6^2} = \frac{5.55 \times 129600}{36} = 1.99 \times 10^4$$

$$H = \frac{L}{N} = \frac{10\, m}{1.99 \times 10^4} = 0.50\, mm$$

9.5.2 Resolution

Clearly, separation between neighbouring peaks, which dictate the magnitude of resolution can be well established by dividing (Δt_r) with the average peak width (w_{av}) measured at the base, as in Figure 9.13:

$$Resolution = \frac{\Delta t_r}{w_{av}} = \frac{0.589\, \Delta t_r}{w_{1/2av}}$$

$$(9.3)$$

In expression 9.3, $w_{1/2av}$ represents the average width at half-height. The more clearly separation between neighbouring peaks, the better the resolution. Figure 9.13 tries to show two peaks with varying resolutions, one having 0.50 and one 1.00 respectively. The dashed lines in Figure 9.13 try to show boundary for individual peaks, and solid lines show the sum of two peaks. The one with resolution of 1.00 has clear peak separation as opposed to that with 0.50.

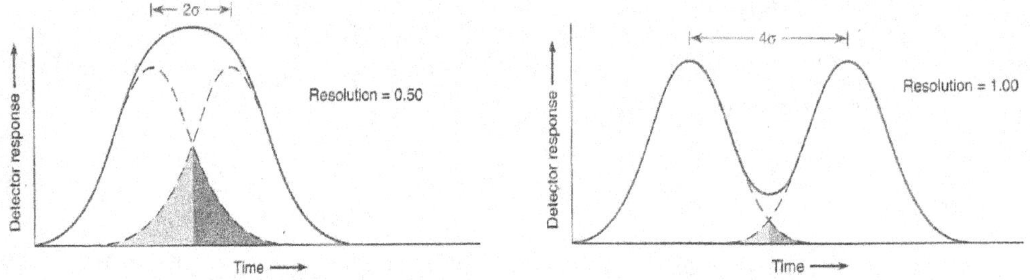

Figure 9.13: Resolution of Gaussian Peaks of Equal Area and Amplitude

9.5.3 Scaling up Separation

At laboratory level analytical chromatography is carried out on a small scale for separation, identification, and measurement of components in the mixture. Conversely, preparative chromatography is often carried out on a large scale to isolate a reasonable quantity of one or more components of a mixture. Typically, analytical chromatography uses long, thin columns for the purpose of obtaining good resolution. On the contrary, preparative chromatography usually uses short, fat columns that can accommodate larger quantities of material, but ending up with relatively poor resolution.

Example 9.2
Suppose you have developed a chromatographic procedure with intention of separating 2 mg of a mixture on a column with a diameter of 1.0 cm, what size of column should you use to separate 20 mg of the mixture? What looks to be straightforward in scaling up is to maintain a column length and increase cross-sectional area to maintain constant ratio of unknown to column volume. It has been established that cross-sectional area is proportional to the square of the column radius (r); therefore, scaling equation is as follows:

$$\frac{large\ load\ (g)}{small\ load\ (g)} = \left(\frac{\text{large column radius}}{\text{small column radius}}\right)^2$$

$$\frac{20\ mg}{2\ mg} = \left(\frac{\text{large column radius}}{0.50\ cm}\right)^2$$

large column radius = 1.58 cm

Example 9.3
(a) Use a ruler to measure retention time and width at the base (w) of octane and nonane in Figure 9.13 to the nearest 0.1 mm.
(b) Calculate the number of theoretical plates for octane and nonane.
(c) If the column is 1.00 m long, find the plate height for octane and nonane.
(d) Use your measurements from part (a) to calculate the resolution between octane and nonane.
(e) Suppose that the sample size for Figure 9.13 was 3.0 mg, the column dimensions were 4.0 diameters x 1.00 m length, and the flow rate was 7.0 mL/min. What size of a column and flow rate should be used to obtain the quality of separation of 27.0 mg of sample?

Solution:
Your answer in part (a) may differ slightly from the one obtained in Figur 9.13, which was measured on a larger figure than one measured in Figure 9.13. Nevertheless, your answer to parts (b) through (e) should be the same as in the above example.
(a) t_r: octane, 200 units; nonane, 260 units
w: octane, 40 units; nonane, 52 units

(b) octane: $N = \dfrac{5.55 \times 200^2}{40^2} = 139$ $\qquad N = \dfrac{5.55 \times 200^2}{40^2} = 139$

Nonane: $\quad N = \dfrac{5.55 \ x \ 260^{2}}{52^{2}} = 139$

(c) octane: $H = \dfrac{L}{N} = \dfrac{1.00 \ m}{139} = 7.2 \ mm$

    nonane : $H = \dfrac{L}{N} = \dfrac{1.00 \ m}{139} = 7.2 \ mm$

(d) resolution $= \dfrac{\Delta t_{r}}{W_{av}} = \dfrac{260 \ - \ 200}{\dfrac{1}{2}(40 \ + \ 52 \ )} = 1.30$

(e) $\quad \dfrac{large \ load \ (g)}{small \ load \ (g)} = \left(\dfrac{large \ column \ radius}{small \ column \ radius}\right)^{2}$

$$\dfrac{27.0 \ mg}{3.0 \ mg} = \left(\dfrac{large \ column \ radius}{2.0 \ mm}\right)^{2}$$

Large column = 6.0 mm, large rate should be proportional to the cross-sectional area of the column, which is proportional to the square of the radius.

$$\dfrac{large \ load \ (g)}{small \ load \ (g)} = \left(\dfrac{large \ column \ radius}{small \ column \ radius}\right)^{2}$$

$$\left(\dfrac{6.0 \ mm}{2.0 \ mm}\right)^{2} = 9.0$$

If the small column flow rate is 7.0 mL/min, the large column flow rate should be (9.0) x (7.0 mL/min) = 63 mL/min.

9.5.4   Band Diffusion

An extremely narrow band of solute that is stationary inside the column slowly broadens due to the fact that the solute molecules tend to diffuse away from the center of the band in both directions. This process is called *longitudinal diffusion*, which begins at the moment solute is injected into the column. The longer the band has traveled in chromatography, the more time it has had to diffuse and the broader it becomes (Figure 9.14). Furthermore, the faster the flow rate, the less time a band spends in the column and the less time there is for diffusion to happen. It is apparent that broadening by longitudinal diffusion is inversely proportional to flow rate. Broadening by longitudinal diffusion can be summarized as in expression 9.4.

$brodening \propto \dfrac{1}{u}$

$$\tag{9.4}$$

whereas u is flow rate

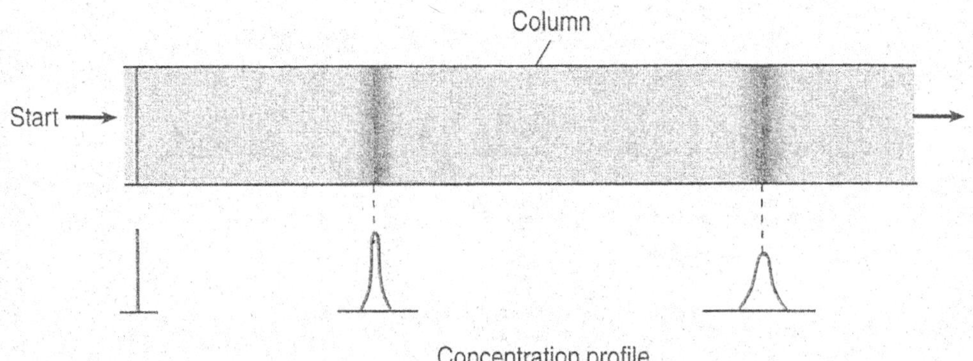

Figure 9.14: Broadening of an Initially Sharp Band of Solute Moving Through a Chromatography Column

9.5.5   Time Required for Solute to Equilibrate Between Phases

Suppose a solute is distributed between the mobile and stationary phases at a given point in time at some location in a column with zero flow rates. Afterward, this automatically will allow the flow to be in motion. Lack of quick reaching of equilibration between the phases results in the solute in stationary phase lagging behind the solute in mobile phase. This broadening is due to the finite rate of mass transfer between phases, which tends to worsen as the flow rate increases.

9.5.6   Optimum Flow Rate in Separation

One should be careful when trying to separate closely spaced bands in order to minimize band broadening in chromatography. The challenge remains for bands, which are too broad because they are always difficult to resolve from one another. For the case of longitudinal diffusion, it decreases with increasing flow rate as shown in expression 9.4, but broadening by the finite rate of mass transfer increases with increasing flow rate see equation 9.5. In fact, there is always an intermediate flow rate that gives minimum broadening and optimum resolution.

In this case, both scientific experience and art of chromatography should be employed to establish conditions such as flow rate and solvent composition to obtain the best separation between mixtures. It should be clearly noted that the rate of mass transfer between phases increases with temperature. Thus, raising the column temperature has an effect in improving the resolution or allows faster separation without reducing resolution.

Broadening by finite rate of mass transfer is expressed as:
$$\text{brodening} \propto u \tag{9.5}$$

9.5.7 Exceptions to Some Band Broadening

Over time, it has been experienced that some mechanisms of band broadening are independent of flow rate. This is best demonstrated in Figure 9.15, which shows a mechanism that is called *multiple paths* and can happen in any column packed with solid particles. Based on the existence of random flow paths which are longer than others, it has resulted in solute molecules entering the column at the same time on the left as are eluted at different times on the right.

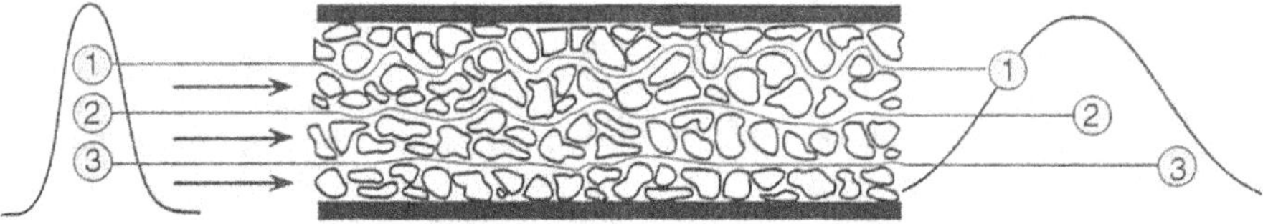

Figure 9.15: Band Spreading From Multiple Flow Path

9.5.8 Plate Height Equation

The three band-broadening mechanisms discussed above in section 9.4.7 are summarized by the *van Deemter equation,* which shows the relationship of plate height (H) and flow rate (u).

$$H = A + \frac{B}{u} + Cu$$

$$(9.6)$$

From plate height equation 9.6 A, B, and C are constants established by the column, stationary phase, mobile phase, and temperature. Indeed, from the equation A represents multiple paths, B/u represents longitudinal diffusion, and Cu represents equilibration time. The equation describes the broadening of a band as it passes through a chromatography column. This is more particularly for the bands with finite width when it is applied to the column; the eluted band emerging from the other end will be broader than predicted by the van Deemter equation. There is always a chance of bands broadening outside the column if there is too much tubing to flow through or if it happens that the detector has relatively large volume. The chromatography will often provide the best results if the lengths and diameters of all tubing outside the column are kept at a minimum. This means there should be no unused spaces inside the column where mixing usually occurs.

9.5.9 Open Tubular Columns

In comparison, a packed column is filled with solid particles coated with stationary phase, and an open tubular column is a hollow capillary whose inner wall is coated with thin layer of stationary phase: Figure 9.17. An open tubular column is free from *multiple paths* broadening, due to the fact that there are no particles in the flow path through stationary phase. Thus, a given length of open tubular column often gives better resolution than the same length of packed column. This means that open tubular column has more theoretical plates or a smaller plate height than the packed column.

As is always the case, the packed gas chromatography column has greater resistance to gas flow than does an open tubular column. Therefore, an open tubular column can be made relatively longer than a packed column with the same operating pressure. Because of its resistance to gas flow, in practice a packed gas chromatography column is often about 2 to 3 m in length, whereas an open tubular column can reach to 100 m in length. The greater length and smaller plate height of the open tubular column provides relatively better resolution than a packed column does. Unfortunately, an open tubular column cannot accommodate as much solute as a packed column, because there is less stationary phase in the open tubular column. Thus, open tubular columns are commonly used in analytical separations, but not for preparative separations.

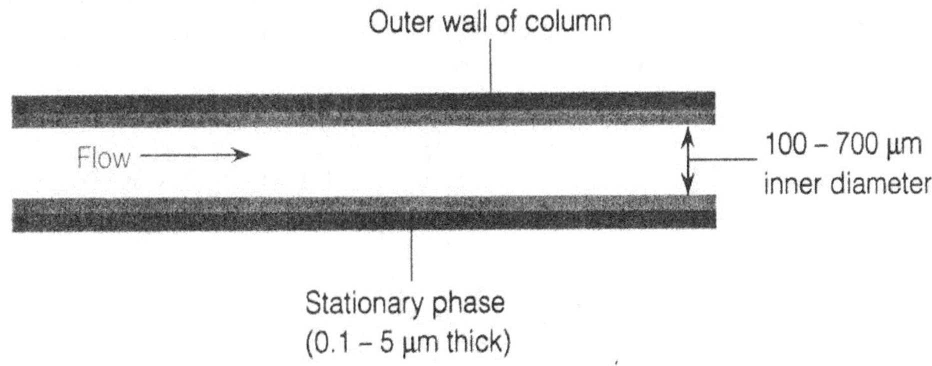

Figure 9.16: Typical Open Tubular Column for Gas

9.6   Capillary Electrophoresis

In simple terms, the migration of ions (cations and anions) in an electric field is called *electrophoresis*. While cations are attracted to the cathode, anions are attracted to the anode. It is apparent that different ions often migrate at different speeds. Capillary electrophoresis is an extremely high resolution (see 9.4.2) separation technique conducted with solution of ions in a narrow capillary tube. When this is carefully done, the technique can allow separation of neutral analytes as well as ions. The technique is capable of separating both macromolecules, such as proteins and Deoxyribonucleic Acid (DNA), and relatively small species, such as K^+ and benzene. Indeed, the technique has capability to measure and separates ions on the basis of their different migration rates in an electric field.

Capillary electrophoresis offers extremely narrow bands, as you can see in the van Deemter equation section 9.4.8, which deals with three mechanisms of band broadening in chromatography, which are multiple flow paths around particles (A in van Deemter equation), longitudinal diffusion (B in van Deemter equation), and finite rate of mass transfer between the stationary and mobile phases (C in van Deemter equation). Open tubular in electrophoresis or chromatography reduces band broadening compared to a packed column. In addition, capillary electrophoresis reduces broadening by removing mass transfer problem (the C term) because it has no stationary phase. The common source of broadening in this case is from longitudinal diffusion of solute as it migrates through the capillary. In actual fact, capillary electrophoresis when used routinely can achieve between 50,000 and 500,000 theoretical plates (see section 9.4.1), which is relatively higher than in chromatography. This can be used in both research laboratories for isolating new proteins and in clinical laboratories for determining protein concentration in blood serum.

9.6.1   How Capillary Electrophoresis Works

This technique involves two processes called electrophoresis and electroosmosis. As we have seen, electrophoresis is the migration of ions in an electric field. On the contrary, electroosmosis pumps all solution through the capillary from the anode towards cathode. The cations in Figure 9.17 migrate from the injection end at the left toward the detection end at the right. Electroosmosis has the ability to sweep both cations, and anions from left to right. The cations seemed to arrive at the detector before the anions.

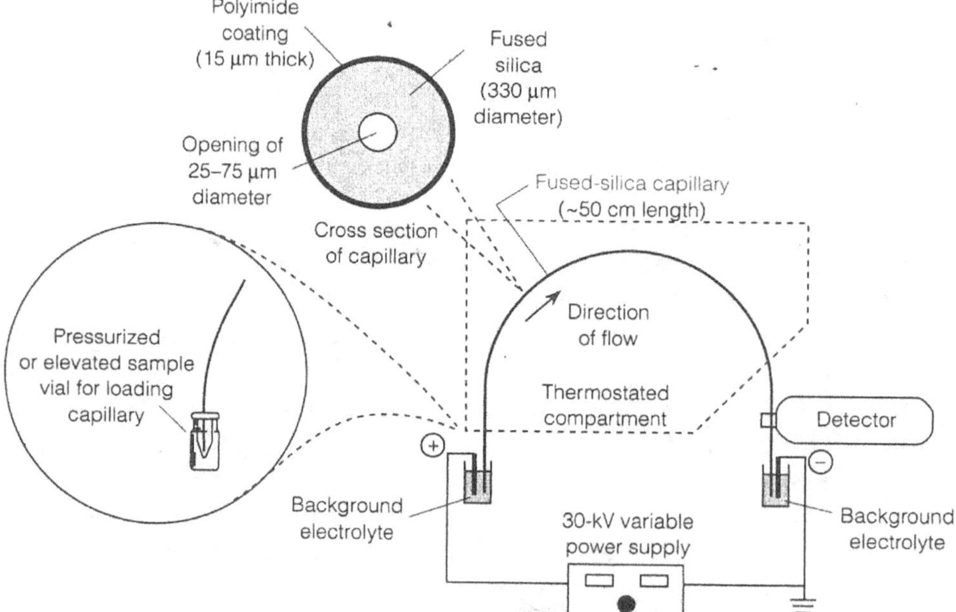

Figure 9.17: How Sample is Injected in the Capillary Electrophoresis

Example 9.4
Which mechanisms of band broadening that operate in chromatography are absent in capillary chromatography?

Solution:
Capillary electrophores eliminate peak broadening from (i) the finite rate of mass transfer between the mobile and stationary phases, and (ii) multiple flow path around the stationary phase particles.

9.6.2   Electroosmosis

The process of pumping fluid inside a fused-silica capillary from the anode toward the cathode caused by the applied electric field is called electroosmosis. For the purpose of clear understanding of electroosmosis, let us look at what happens inside the wall of the capillary. As can be seen in Figure 9.18 (a), the wall is covered with silanol (Si-OH) groups that are negatively charged (Si-O) above pH 2. In the latter figure the capillary wall and solution is immediately adjacent to the wall form an electric double layer. The electric double layer is composed of (1) positive charges in solution adjacent to the wall, and (2) an equal negative charge fixed to the wall. Upon applying electric field, cations will be attracted to the cathode and anions are attracted to the anode. The diffused part of the double layer with excess cations drives the rest of the solution in the capillary toward the cathode, as can be seen in Figure 9.18 (b). The flow speed depends on the strength of the applied electric field.

Electroosmotic flow is often uniform across the diameter of the liquid, as can be seen in Figure 9.18 (c) because it is driven by ions in the walls of the capillary. It should be clearly noted that the chance of moving band broadening is due to diffusion. Typically, electroosmosis decreases at low pH because the wall loses its negative charge when Si-O- is converted to Si-OH and the number of cations in the double layer disappears. Through addition of an ultraviolet absorbing neutral solute, for example methanol, to the sample and measuring the time taken (migration time) to reach the detector it can be achieved and is called electroosmotic velocity.

(a)

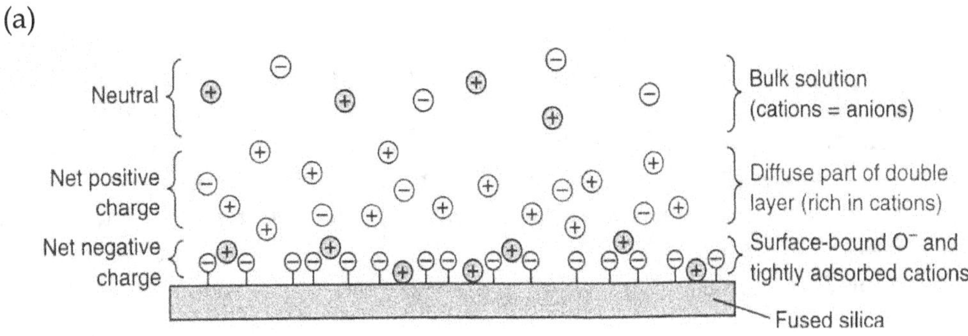

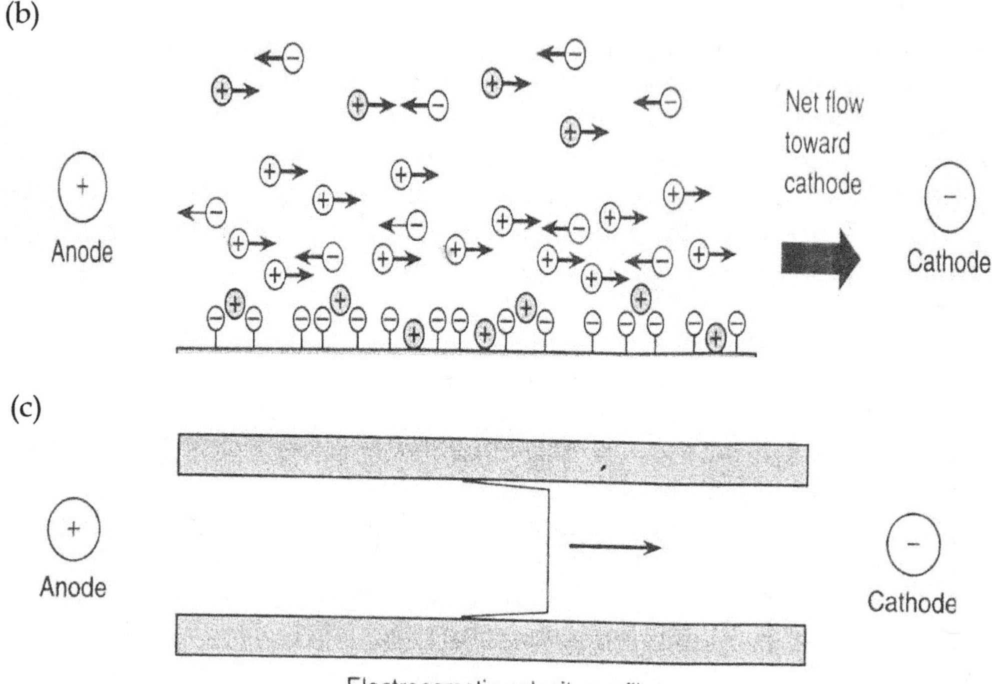

Figure 9.18: Electric Double Layer and Electroosmotic Velocity Profile

From observation of both Figure 9.17 and 9.18, electroosmosis is from left to right due to the reason that the cations in the double layer are attracted to the cathode. It should be noted that at neutral or high pH, often elecroosmosis is much faster than electrophoresis and the overall flow of anions is heading to the right. However, at low pH, electroosmosis is relatively weak and anions may start to flow toward the left and as a result fail to reach the detector. At low pH, one can separate anions by reversing the polarity, which will affect the sample side to be temporarily negative and the detector side positive. At the moment the electrophoresis discussed is generally called capillary zone electrophoresis, in which separation is based on variation of electrophoretic velocity of different ions. It is a good idea to look at another important type of electrophoresis that can separate DNA, which is capillary gel electrophoresis.

9.6.3   Capillary Gel Electrophoresis

Macromolecules are easily separated in capillary gel electrophoresis through sieving as they migrate through a gel inside a capillary tube. Contrary to large molecules in molecular exclusion chromatography (9.2 iii), the small molecules in gel electrophoresis travel much faster than the large molecules through the gel. On the other hand, the large molecules become entangled, which results in reducing their motion. The main function of capillary gel electrophoresis is used in sequencing the human *genome*. At most, 500 different lengths of DNA are separated in less than 20 minutes by this technique whereby the shortest DNA migrates faster.

9.7 Problems

9.7.1. What is the difference between eluent and eluate?

9.7.2. Which column gives narrower bands in a chromatographic separation: plate height = 0.2 mm or 2 mm? Provide a short explanation to support your choice.

9.7.3. Explain why a chromatographic separation normally has an optimum flow rate that gives the best separation.

9.7.4. (a) Why is longitudinal diffusion a more serious problem in gas chromatography than in liquid chromatography?

(b) Give your opinion on why the optimum linear flow rate is much higher in gas chromatography than liquid chromatography.

9.7.5. Provide clear explanations on the following:

(a) Why does an open tubular gas chromatography column give better resolution than that of the same length of packing column?

(b) Why is it desirable to use a very long open tubular column? What difference is there between packed and open tubular columns?

9.7.6. (a) How do many theoretical plates produce a chromatography peak eluting at 12.83 minutes with a width at half height of 8.7 seconds?

(b) The length of the column is 15.8 cm. Find the plate height.

9.7.7. Two components of a 12 mg sample are adequately separated by chromatography through a column of 1.5 cm in diameter and 25 cm long at a flow rate of 0.8 mL/min. What size of a column and what flow rate should be used to obtain similar separation with a 250 mg sample?

9.7.8. A chromatogram in the figure below has a peak for isooctane at 13.81 min. The column is 30.0 m long.

(a) Measure $w_{1/2}$ and calculate the number of theoretical plates for this peak.

(b) Find the plate height

(c) The ratio of $w/w_{1/2}$ from Figure 9.13 for a Gaussian peak is $4\sigma/2.35\sigma = 1.70$. Measure the width at the base for isooctane in the chromatogram of this problem. Compare the measured ratio $w/w_{1/2}$ to the theoretical ratio.

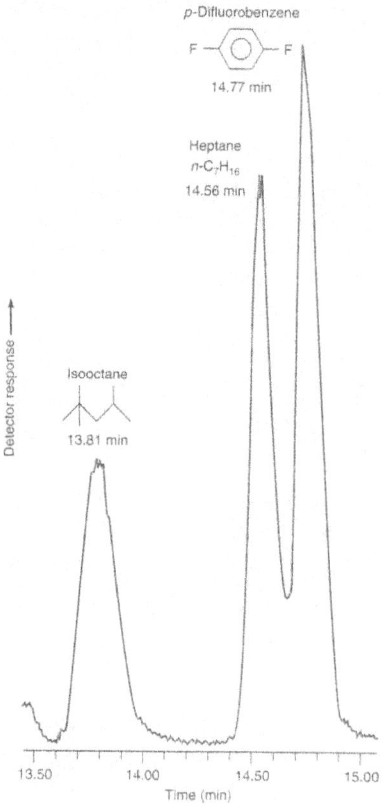

9.7.9. Consider the peaks for heptane (14.56 min) and *p*-difluorobenzene (14.77 min) in the chromatogram for problem 9-8. The column is 30.0 m long.

(a) Calculate the number of plates and height for the two compounds

(b) Measure w for each peak and calculate the resolution between the two peaks.

9.7.10. A column with 3.0 cm in diameter and 32.6 cm long gives adequate resolution of a 72.4 mg mixture of unknowns initially dissolved in 0.50 mL.

(a) If you wished to scale down to 10.0 mg of the same mixture with minimum use of chromatographic stationary phase and solvent, what length and diameter column would you use?

(b) In what volume would you dissolve the sample?

(c) If the flow rate in the large column is 1.85 mL/min, what should be the flow rate in the small column?

9.7.11. (a) When you are trying to separate an unknown mixture by reversed-phase chromatography with 50% acetonitrile-50% water, the peaks are eluted between 2 and 3 minutes and they are too close together to be well resolved for quantitative analysis. Should you use higher or lower percentage of acetonotrile in the next run?

(b) Suppose that you are trying to separate an unknown mixture by normal-phase chromatography with the solvent mixture 50% hexane-50% methyl *t*-butyl ether (which is more polar than hexane). The peaks are too close together and are eluted rapidly. Should you use a higher or lower percentage of hexane in the next run?

9.7.12. (a) The plate height in a particular packed gas chromatography column is characterized by the *van Deemter* equation where $A = 1.50$ mm $B = 25.0$ mm. mL/min, $C = 0.025$ mm. min/mL, and u is flow rate in mL/min. Construct a graph of plate height versus flow rate and find the optimum flow rate for minimum plate height.

(b) In the *van Deemter* equation, B is proportional to the rate of longitudinal diffusion. Predict whether the optimum flow rate would increase or decrease if the rate of longitudinal diffusion were doubled. To confirm your prediction, increase B to 50.00 mL/min, construct a new graph, and calculate the optimum plate height increase or decrease.

9.7.13. (a) Capillary electrophoresis was conducted with a solution whose pH was 9, at which the electroosmotic velocity is greater than the electrophoretic velocity. Draw a picture of the capillary, showing the placement of anode, cathode, injector, and detector. Show the direction of net flow for each ion.

(b) If the pH is reduced to 3, the electroosmotic velocity is less than the electrophoretic velocity. What are possible directions cations and anions will migrate to?

9.7.14. A molecular exclusion column has a diameter of 7.8 mm, and a length of 0.30 cm. the solid portion of the particles occupies 20% of the volume, the pores occupy 40%, and the volume between particles occupies 40%.

(a) At what volume would totally excluded molecules be expected to emerge?

(b) At what volume would the smallest molecules be expected?

(c) A mixture of polyethylene glycols of various molecular masses is eluted between 23 and 27 mL. What does this imply about the retention mechanism for these solutes on the column?

9.8. References

Arakawa, T., Ejima, D. Li, T. and Philo, S. J. (2010). The Critical Role of Mobile Phase Composition in Size Exclusion Chromatography of Protein Pharmaceuticalas, *J. Pham. Sci. 99*, 1674.

Fekete, S., Beck, A. and Venthey, J. L. (2014). Theory and Practice of Size Exclusion Chromatography for the Analysis of Protein Aggregates, *J. Pham. Biomed. Anal. 101*, 161.

Ferguson, G. K. (1998)."Quantitative HPLC Analysis of an Analgesic/Caffeine Formulation: Determination of Caffein," *J. Chem. Educ. 75*, 467.

Grob, K. (2001). *Split and Splitless Injection for Quantitative Gas Chromatography*, Wiley, New York.

Hanai, T. and Smith, R. M. (1999). *HPLC: Practical Guide* Springer-Verlag, New York.

He, Y., Friese, O. V., Schlittler, M. R., Wang, Q., Yang, X. and Bass, L. A. (2012). On-line Coupling of Size Exclusion Chromatography with Mixed-Mode Liquid Chromatography for Comprehe, *J. Chromatogr. A.*, *1262*, 122.

Hinswan, J. V. and Ettre, L. S. (1994). *Introduction to Open Tubular Gas Chromatography*, Advanstar Communications Cleveland, OH.

Jennings, W., Mittlefehldt, E. and Stremple, P. (1997). *Analytical Gas Chromatography*, 2nd ed. Academic Press, SanDiego.

Lucy, C. A., Baryla, N. E. and Yeung, K. K. C. (2004). *Capillary Electrophoresis of proteins and Peptides*, M. L. Strege (Ed.) Human Press.

Maxwell, E. J. and Chen, D. D. Y. (2008). Twenty Years of Interface Development for Capillary Electropheresis Electrospray Ionization-Massspectrometry, *Anal. Chim. Acta*, *627*, 25.

Meyer, H. T. . (1999). *Practical High-Performance Liquid Chromatography*, 3rd ed. Wiley, New York.

Poole, C. F. (2003). *The Essence of Chromatography*, Elsevier, Amsterdam.

Remcho, V. T., McNair, H. M. and Rasmussein, H. T. (1992). "*HPLC* Method Development with the Photodiodre Array Detector" *J. Chem. Ed.* 69, A117.

Torto, N., Mmuacefe, L. C., Jonas, F., Nkoana, B., Chimuka, L., Nindi, M. M. and Ogunfowolean, A. O. (2007). Sample Preparation for Chromatography: *An African Perspective, 1153*, 1.

Zhu, Z., Joann, J. and Liu, S. (2012). Protein Separation by Capillary Gel Electrophoresis: A Review, *Anal. Chim Acta*, *769*, 21.

Zwir-Ference, A. and Biziuk, M. (2006). Solid Phase Extraction Techniques-Trends, Opportunities and Applications, *J. of Environ. Stud. 15*, 677.

Chapter **10**

ANALYSIS USING BIOCHEMICAL REACTIVITY

10.1　Introduction

The characteristic to be considered for good differentiating quality is the extent to which it can differentiate the intended analysis from the rest of the sample. The more powerful the discriminating characteristic is, the less interference and the lower the detection limit. In actual fact, the detection limit in a chemical analysis is often set by the relative amount of interfering species. This means the detection limit has something to do with the selectivity of differentiating characteristic than the sensitivity of the used measurement system.

So far, the chemical reactions we have seen can effectively distinguish one type of compound from another; for example, those which can react with an acid from those which do not. We have seen how chromatographic methods have a great potential in discriminating capacity in various circumstances; however, as the concentration to be analyzed gets lower, the greater must be the extent of the selectivity. The exceptional aspect of the chemical reactions that occur in biological systems is their unbelievable extent of selectivity, which is due to the limited number they can react with.

The most commonly encountered biochemical reactions in chemical analysis include enzyme reactions and antigen-antibody reactions. The latter reactions have been used in quantitatition, detection, and separation in a creative manner.

10.2　Enzyme Reactions

Enzymes are relatively large proteins that serve as catalyst for biological reactions systems. They generally differ from inorganic catalysts in two specific crucial aspects. First, enzymes have relatively large and more complex structures than inorganic catalysts, with molecular weight ranging between 10^4 to greater than 10^6 amu. Second, each enzyme works for a specific type of reaction more than an inorganic catalyst, usually catalyzing the specific single reaction of a single compound, which is called *enzyme's substrate*. This is where the enzymes are attached to the chemical function they perform. The enzyme *amylase* found in the human digestive system is used as an example in expression 10.1. It catalyzes the breakdown of starch to yield glucose, but has no effect on cellulose, despite the fact that starch and cellulose are structurally similar. That is why humans can digest potatoes (starch), but not grass (cellulose).

197

How enzyme activity is often measured? The catalytic activity of enzyme is often measured by its *turnover number*. The latter means the number of substrate molecules acted on by one molecule of enzyme per second. Most enzymes have turnover numbers which range between 1 and 20,000, but few have much higher values. For example, carbonic anhydrase, which catalyzes the reaction of CO_2 with water to form HCO_3^- ion, acts on 600,000 substrates per second.

Enzymes usually work like the lock-and key model. Thus, the enzyme is pictured as an irregularly shaped molecule in Figure 10.1 with a crevice in the middle part. The active site is located inside the crevice, a small portion with the shape and chemical composition necessary to bind the substrate and catalyze the reaction. In fact, the active site acts like a lock into which only a specific key (substrate) whose shape and structure are complementary to those of active site can fit. An enzyme's active site is made by various acidic, basic, and neutral amino acids inside chains, all well positioned for maximum interaction with the substrate. In actual fact, there are no covalent bonds which exist between enzyme and substrate, but rather they are held together only by hydrogen bonds and weak intermolecular attractions. The enzyme and substrate which are positioned in a well-defined arrangement of atoms in the active site facilitate the reaction of substrate molecule, and enzyme plus product separation. More reactions examples facilitated by enzymes include these in expression 10.2

All enzymes indicated in expression 10.2 have the suffix "ase" which usually indicates enzyme names and the chemical function they perform, and often the name includes the molecule for which they facilitate the reaction. For conversion of alcohol to an aldehyde the dehydrogenase enzyme requires another reactant (NAD^+) to enable such a conversion to happen. The reduction of NAD^+ (nicotinamide adenine dinucleotide) by addition of H atom results in NADH. Under this circumstance, NAD^+ is called a *coenzyme* or a *cofactor*. The latter is often reoxidised to NAD^+ by a separate enzymatic reaction to finish the cycle and to reconstitute the dehydrogenation reaction catalysts.

In enzyme catalyzed reaction the catalyst provides a lower energy path for the reaction to occur in very fast reaction rate. For the purpose of appreciation of the extent of enhancement achieved by enzymes, one can think of the reaction of *fumarate* (the basic form of *fumaric acid*) with water, hereunder in expression 10.3. As can be seen, water added across the double bond forms *malate* (the basic form of malic acid). It has been noted that such a reaction in neutral solution with the enzyme catalyst fumarase is 3.5×10^{15} times faster than if the catalyst was not used. This means the product formed in one minute in the presence of fumarase present would require 6.6 billion years to form without it.

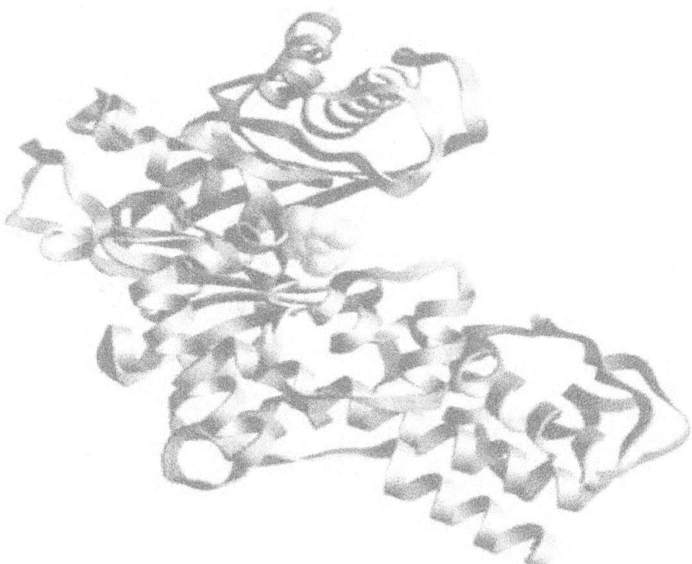

Figure 10.1: Active Site of the Enzyme Hexose Kinase

The enzyme reactions attain specificity through complex formation, which is adversely affected by conformation. In order for an oxidase to catalyze the oxidation of substrate molecule, the substrate must complex with oxidase at specific location in the enzyme called *active site*. This is three dimensional holes in the enzyme protein. Complication will always require appropriate orientation corresponding to atoms in the active site of the oxidase.

10.2.1 The Effect of pH on Enzyme

Enzymes are always sensitive to the pH of the environment they are found in. Obviously, amino acids, which enzyme proteins are made of, are weak acids and have different forms at both low and high pH values. The most disturbing aspect regarding pH dependency is the *denaturation* of the protein that occurs in very low pH and very high pH solutions. Denaturation is a complete loss of tertiary and quaternary structure. This will lead to entire loss of enzyme activity. Change in pH has an effect on the charge state of acidic species in the active sites. The pH also has an effect on the conjugate form of amino acid that can affect the tertiary structure of the peptide and conformation of active site. Lastly, the pH has an effect on the extent of activation or inhibition reactions.

Most enzymes function at physiological pH which is about 7.4 and are expected to have significant activity at that pH, as can be seen in the general form of function shown in Figure 10.2. When using enzymatic reaction in the laboratory, the activity of the enzyme should be controlled by use of buffer solution in optimal pH range. Clearly, it should be noted that a good buffer to be used should be one with a conjugate acid-base pair with a pK'_a close to the desired solution pH. For sure enzymatic reactions which require the greatest buffers that span the useful range have been developed by biochemists, as can be seen in Table 10.1.

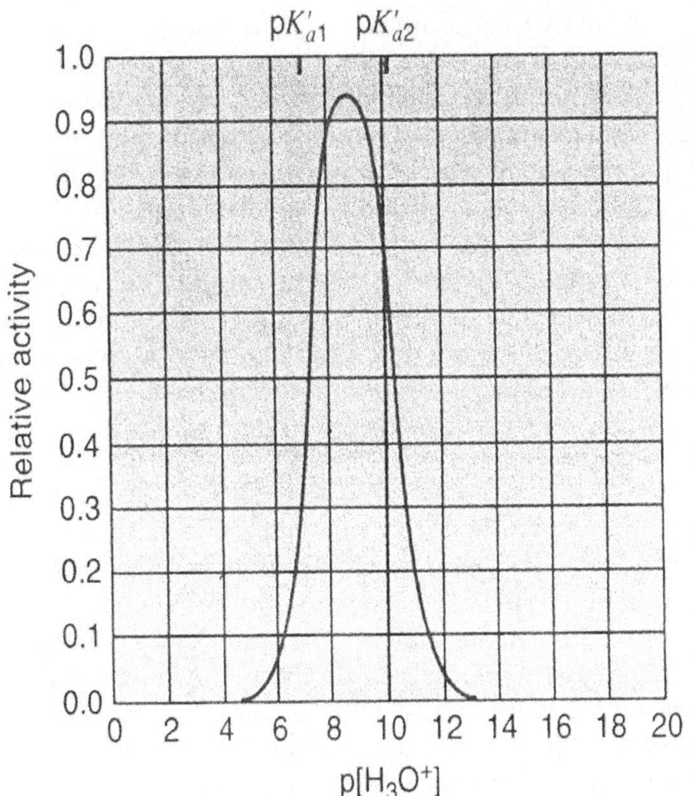

Figure 10.2: Relative Activity of a Typical Enzyme Versus pH (12.1)

Table 10.1: Some Buffers Used with Biochemical Reactions

Abbreviation	Acid	pK'a
MES	2-(N-Morpholino)ethanesulfonic acid	6.1
ADA	N-(2-Acetamido)iminodiacetic acid	6.6
PIPES	Piperazine-N,N'-bis(2-ethanesulfonic acid)	6.8
ACES	N-(2-Acetamido)-2-aminoethanesulfonic acid	6.8
MOPS	(3-N-Mopholino)-propanesulfonic acid	7.2
HEPES	N-2-Hydroxyethylpiperazine-N'-2-ethanesulfonic acid	7.5
HEPPSO	N-2-Hydroxyethylpiperazine-N'-3-propanesulfonic acid	8.0
TRIS	Tris(hydroxymethyl)aminomethane	8.1
Bicine	N,N-Bis(2-hydroxyethyl)glycine	8.3

10.3   Antigen and Antibodies

The proteins which are produced by an animal to protect itself against the activity of foreign substances introduced into its tissue are called *antigens*. The latter respond through binding to the large molecules in the foreign substance for the purpose of destroying, deactivating or removing them. As is always the case for enzyme, each *antibodiy* generated is for a specific intruder molecule. In actual fact, each antibody has been designed and produced for a specific invader. The molecule that activates the production of an antibody is called *parathion antigen* or an *immunogen*. In terms of structure, antigens are very large molecules like proteins, oligonucleotides or polysaccharides. The portion of antigen molecule, which activates the production of the antibody is called the *haptenic determinant*. This often contains an organic functional group such as dinitrophenyl that is important for the antibody triggering process.

As we have seen, the bonding force between the enzyme and substrate is exactly the same as the bonding force between the antibody and the antigen. This means the bond is a combination of hydrogen bonding, van der Waals forces, coulombic and hydrophobic interactions. Because both antibondies and antigens are very large molecules, just portions of the two are often involved in such bonding. These portions are called *epitope* (for antigen) and *paratope* (for the antibody). Experiments have shown that the bonding between antibody-antigen complex is reversible and generally represented as in expression 10.4, where Ab and Ag are antibody and antigen, respectively.

$$Ab \quad + \quad Ag \quad \rightleftharpoons \quad AbAg \tag{10.4}$$

This complex has generally quite a large formation constant (of the order 10^8 to 10^{10}). The formation constant is established in the same ways as we have seen in chapter six, section 6.2 and it is very specific. The large area of overlap between the epitope and paratope contribute to specificity of the formation constant. This means that the tendency of antibody-antigen reaction is a highly differentiating characteristic and therefore makes it of great analytical value.

10.4   Separation with Biochemical Reaction

In biomolecule reactions there are two important aspects for separation. One can carry out separation of the products from the reaction mixtures when that is a necessary part of the determination. In addition, it is useful for selective bonding of enzyme and antibondies, which facilitate the separation of various molecules based on the extent of bonding on the molecules. The great mass and size of the proteins involved in biochemical reaction has made such techniques based on these physical characteristics more useful.

10.5 Separation Based on Size

Filtration using a filter paper with the pore of a given size to differentiate the materials is the simplest separation technique. Generally, the size ranges of the ions, molecules, and particles of interest can be seen in Table10.2.

Table 10. 2: Particles Sizes

Particle	Diameter , nm
Small molecules and ions	< 1.0
Proteins, viruses	1.0 - 100
Bacteria	100-1000

A good example in this case is *cellophane*, which is a thin film of cellulose acetate with pore size ranging to 4-8 nm. Macromolecules with molecular weight greater than 10,000 would roughly have 2 nm diameters, which make it difficult to pass through *cellophane*. When a cellophane membrane is used to separate solvent from electrolyte ions, which will pass freely through it, the proteins will not pass through. It should be noted that the driving force of this separation is the concentration differences between two solutions; the process is called *dialysis*. Consider a volume of solution containing proteins and an undesirable electrolyte placed in a dialysis container, which in turn is placed in a beaker with relatively larger volume of the desired solvent buffer, as can be seen in Figure 10.3.

After a few hours have elapsed, the buffer and electrolyte concentrations will have reached equilibrium, removing most of unwanted electrolye and replacing them with the desired buffer solutions. In case you want to improve the solvent composition much more, the process can be repeated with fresh buffer.

What often happens in filtration is that a solution is forced through the filter material. Only small particles and solvent smaller than the pore sizes will go through; the larger ions and molecules remain on the inlet side of the filter material. Chemists have developed a variety of polymeric material with varying pore size for this purpose. Since the pores are necessarily very small, the force of gravity is not sufficient enough to drive the solution through the membrane at a high rate. When a pump is employed to supply the force of gravity, the process is called *ultrafitration*.

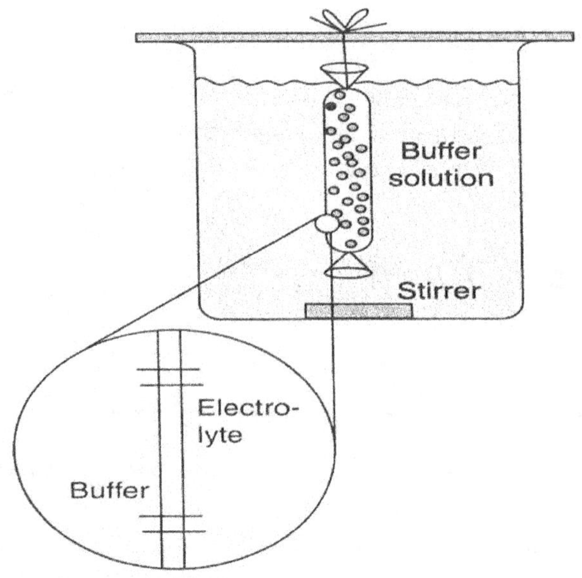

Figure 10. 3: Process of Dialysis, Electrolytes, Solvents and Buffer Components

Filtration as a batch process is one which can discriminate between ions and molecules that are much larger or much smaller than the pore size. For the case when much finer degrees of discrimination are required, a continuous process is needed, as we have seen in chapter 9. Indeed, a special type of chromatography has been developed for separation based on particle size. It is called *gel permeation chromatography*. In this method the column is packed with beads of a porous gel. Usually, the pore size of the gel materials is carefully controlled during its manufacturing. The mobile phase is often aqueous buffer solution in which macromolecules are stable.

When part of aqueous mobile phase containing the sample gets in contact with the bead packing, some of ions and molecules that can diffuse into the pores of the bead will diffuse. The degree of diffusion into the pore will depend on both size and shape of the particle relative to the pore dimensions. The relative velocity of the particle passing through the column while it is in the pore is zero; while it is in mobile phase, it has mobile phase velocity. Thus as in the case of partition chromatography in chapter 9, section 9.1 (ii), the relative velocity of the component is the function of time it spends in the stationary phase. However, under this situation there are important limits. One is that those particles which are too large to enter the pores will always remain in the mobile phase, and all will move with mobile phase velocity. On the other hand, particles that are relatively small can readily diffuse through entire pore structure of the beads and will all possess the minimum velocity. What has been noted between these two extremes, is that particles elute in order of size, whereby the largest particles elute first.

10.6   Separation Based on Complexation Reaction

Due to the specific binding of enzymes, antibodies and other proteins, this characteristics can therefore be used as one of the techniques of separation, as can be seen in Figure 10.4. When using this technique, a molecule that binds to the analyte that is to be separated is mobilized on a solid support. This material is afterwards used in a column through which a solution containing the analyte material is passed. Selectively, the analyte will complex with the immobilized ligand and therefore remain attached to the solid support. Thereafter, the rest of the material in the sample is eluted in void volume. This means the space between the beads (*void volume*).

This is followed by changing mobile phase composition to a given formulation which will allow the analyte to unbind from the immobilized ligand, and therefore be eluted in very pure form. This approach is called *affinity chromatography,* as you can see chapter 9.1 (iv) because the separation technique used is based on attraction between the mobile phase and the affinity of the sample. This approach has been commonly used in a technique called Immobilized Metal Affinity Chromatography (IMAC). In the latter approach the binding affinities of the analytes require to be in an immediate range such that the effective capacity factor is not too large or not too small. For a single affinity separation, a high and highly specific binding affinity is always required.

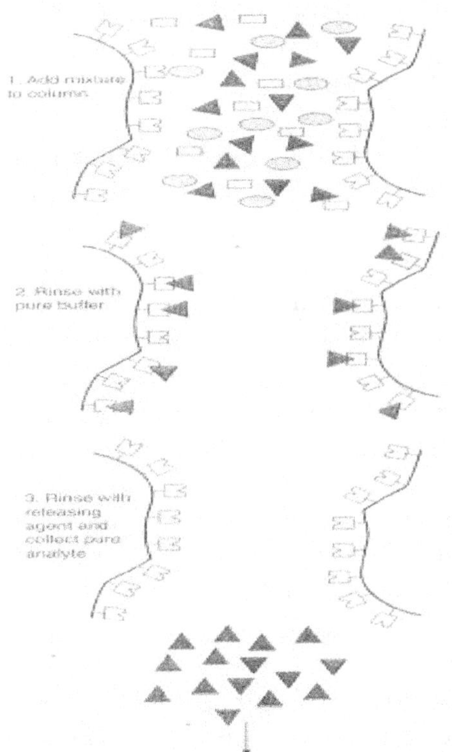

Figure 10.4: Steps Involved in Affinity Chromatography

The other sample components are carefully washed from the column, and then the analyte is released in a single step. This has proved that this technique is a batch rather than continuous process, and thus not strictly chromatography. Surprisingly the analysis of 2,4-dichlorophenoxyacetic acid (the pesticide 2,4-D) in water can be achieved in a pure form of 30,000:1 by affinity chromatography. Regardless of the highest efficiency of this technique, it has two difficulties in implementing it. First, the considered immobilized ligand can still be free to some extent to react with analyte. Second, there is dissociation possibility of the analyte after the rest of the sample has been washed away. Usually, unbinding can be achieved through competitive binding or changing the solution conditions. In the first case, the elution solution contains a molecule that has a higher complex formation constant for the ligand than does the bound analyte. This molecule will easily displace the analyte from the ligand, and allow it to be washed out. A commonly used technique involves change in ionic strength or pH of the solution to distort the active site of the protein, which will lower its formation constant.

The most useful biochemical species that have been used in affinity chromatography include enzymes, antibodies, lectins, receptor, proteins, and nucleic acid. Lectins are a class of proteins that have specific binding to certain carbohydrate groups. Mostly used immobilized species include enzymes, the substrate, coenzyme, competitive inhibitor and the desired enzyme is selectively removed from the sample solution.

Affinity chromatography has a significant role to play in chemical analysis. This technique has at least two advantages, which include high selectivity separation and only a relatively small volume of the sample neededs to be used. The use of this technique has greatly improved the detection limits of many instrumental techniques. It can certainly handle difficult separation of complex mixtures, which include the separation of optical isomers. It should be remebered that scientists are tirelessly continuing to develop new methods and materials for affinity chromatography. Due to the complexity of the fluid found in biological fluids, such specific methods remain essential in dealing with such types of sample.

10.7   Separation Based on Both Mass and Density

The *ultracentrifuge* is commonly used in biochemical research. This technique enables macromolecules to be forced and form sediment roughly in order of their mass, whereby most of massive molecules are forced to settle at the bottom of the centrifuge tube. Such separation is attained by placing the sample solution in special centrifuge tubes which are then placed in special centrifuge test tube holders. These test tubes are centriguged rapidly, whereby the bottom of the tube spins outward from the centre of the rotation. This is happening due to the existing relationship between the centrifugal force on molecule (mass) in the tube and rotational velocity of the centrifuge. This force is counteracted in solution by the need of the molecule to displace an equivalent volume of the solution. This sedimentation force, which is due to rotation of the tube can be calculated as in expression 10.5

$$F_{sed} = (m_m - m_s) w^2 r \tag{10.5}$$

where m_m is the actual mass of molecule, m_s is the mass of equivalent volume of solution, w is rotational speed of centrifuge in radian per second, and r is the distance of the molecule from the centre of rotation in centimeters. The mass of an equivalent volume is equal to the volume of the particle, V_m times the density of the solution ρ_s. The volume of the molecule can be calculated by multiplying the mass of the molecule with the *partial specific volume* $\bar{V}_m$ of the molecule in cube centimeters per gram. These two equations (10.6 and 10.7) are as follows:

$$m_s = V_m \rho_s \tag{10.6}$$

and

$$V_m = m_m \bar{V}_m \tag{10.7}$$

Upon combining expressions 10.5, 10.6 and 10.7, the sedimentation force which relates the force and molecular mass is obtained as shown in expression 10.8.

$$F_{sed} = m_m (1 - \rho_s \bar{V}_m) w^2 r \tag{10.8}$$

As can be seen in expression 10.9, the product of the solution *density* and the *partial specific volume* of the molecule must be less than 1 for the sedimentation to happen. Nevertheless, this point can be easily perceived if expression 10.8 is rearranged in terms of relative densities of molecules and solution as in expression 10.9. In this expression F_{sed} stands for force of sedimentation, ρ_s sediment density and ρ_m stands for the density of molecule.

$$U_{sed} = m_m (\rho_m - \rho_s) w^2 r \, \bar{V} \tag{10.9}$$

During centrifugation process the sedimentation force will accelerate the molecule toward the end of the tube. From the solution the molecule gains enough velocity, the friction force fv trying to hold this motion back increases. This force is proportional to both the friction coeffient of the molecule f and the velocity v. As the velocity increases to the point where the frictional force is equal to the sedimentation force, the velocity always becomes constant. The expression for the limiting velocity is called the sedimentation rate U_{sed} as can be seen in expression 10.10.

$$U_{sed} = \frac{m_m(1 - \rho_s \bar{V})w^2 r}{f}$$

$$(10.10)$$

The actual mass of molecule m_m in terms of molecular mass of the molecule, MW/N_A (where N_A is Avogadro's number) must be substituted in expression 10.10 for m_m and expression 10.11 is obtained.

$$U_{sed} = \frac{MW_m(1 - \rho_s \bar{V}_m)\, w^2 r}{N_A f}$$

$$(10.11)$$

Obviously, the rate of sedimentation is proportional to $w^2 r$. In fact, this can be measured by noting down the rate at which the species changes with time. This rate is expressed as $U_{sed} = dr/dt$ and is substituted in expression 10.11 (and followed by integration) and expression 10.12 is obtained.

$$\frac{1}{r}\frac{dr}{dt} = \frac{MW_m(1 - \rho_s \bar{V}_m)\, w^2}{N_A f} = \frac{\Delta \ln r}{\Delta t}$$

$$(10.12)$$

where $\Delta \ln r$ is the change in the natural log of position (in centimeters) over the time Δt.

This means the sedimentation rate measured is proportional to w^2. Through rearrangement of the expression 10.12 sedimentation coeffitient **s** can be obtained in expression 10.13 and has unit of seconds: see example 10.1. It is common practice to measure the value of **s** in many biological molecules in the buffer solutions of which are used with them.

$$\frac{MW_m(1 - \rho_s \bar{V}_m)}{N_A f} = \frac{2.303\, \Delta \log_{10} r}{w^2\, \Delta t} = s$$

$$(10.13)$$

Example 10.1
Suppose we wished the sedimentation rate to be such that a protein with an **s** value of 10 x 10⁻¹³ s would move 0.1 cm (from r of 6.0 to 6.1 cm) in 5 hours. Calculate the rotational speed which will be required.

Solution:

$$w^2 = \frac{2.303(\log 6.1 - \log 6.0)}{1.0\ x\ 10^{-12}\ x\ 1.8\ x\ 10^4\ s}$$

$$= 9.2\ x\ 10^5\ s^{-2}$$
$$w = 960\ \text{rps}$$

The rotational speed can be converted to be rotation per minute (rpm) by multiplying rps by 60 to get 57,000 rpm. The latter value is essentially towards the upper end of rotational speeds available in commercial ultracentrifuges.

10.8 Problems

10.8.1. What is an enzyme, and what is a substrate?

10.8.2. Is the enzyme consumed in carrying out the enzyme-catalyzed reaction? Is the cofactor altered by catalyzed reaction?

10.8.3. What is the active site of an enzyme?

10.8.4. List all the species that might have to be bound to an enzyme for the catalytic reaction to occur.

10.8.5. Which of the buffer systems in Table 10.1 would be suitable for maintaining a pH of 7.5? How would you choose among them?

10.8.6. In an animal under invasion by a foreign protein, which comes first, the antibody or the antigen?

10.8.7. What is the nature of the bond formed between the *epitope* and the *paratope* regions of the antibody and the antigen? Which regions belong to which molecules?

10.8.8. Provide your comment on the desirability of having very uniform pore sizes for the materials used in filtration and in *gel permeation* chromatography.

10.8.9. Why does affinity chromatography require an immobilized ligand?

10.8.10. What must be done to release the analyte from immobilized ligand in affinity chromatography?

10.8.11. Are the immobilized ligands in affinity chromatography limited to antigens or antibodies?

10.8.12. Is the serum normally produced in response to an intrusion polyclonal or monoclonal? Why?

10.8.13. The pH of a solution can affect the activity of enzymes in two ways. One is generally reversible; the other may not be. What are the two actions of pH on enzyme activity?

10.8.14. It is desired to completely separate all molecules that have a sedimentation coefficient of 30 S from a homogeneous solution. The ultracentrifuge tubes are 5 cm in length and are filled to within 1 cm of the top. The bottom of the tube is 11 cm from the center of rotation, and the rotational velocity is 50,000 rpm. How long should you leave the centrifuge spinning to ensure complete settling of the desired molecules?

10.8.15. How is the partial specific volume of a macromolecule related to its density?

10.9 References

Dasilva, A. C., Honorato, T. L., Cavalcante, R. S., Franco, T. T. and Rodrigues, S. (2012). Effect of pH and Temperature of Enzyme Activity of Chitosanase Produced Under Solid State Fermentation by *Trichoderma SPP. India Journal of Microbiology, 52,* 60.

Dombrowski, T., Wilson, G. and Thurman, E. (1998). Investigation of Anion-exchange and Immunoaffinity Particle, Anal. Chem. 70, 1969.

Eisenthal, R., Peterson, E. M., Daniel, M. R. and Danson, J. M. (2006). The Thermal Behaviour of Enzymes: Implication for Biotechnology, Trends Biotechnol. 24, 289.

Ghirlando, R. (2011). The Analysis of Molecular Interactions by Sedimentation Equilibrium: Modern Analytical Ultracentrifugation: Methods, 58, 145.

Goodman, T. (2007). Centrifuge Safety and Security, American Laboratory 01.

John, E. (2013). Factors affecting enzyme activity, *ESSAI, 10,* 48.

Lavive, C., Lunte, S., Zhong, M., Perkins, M., Wilson, G., Gokulrangan, G., Williams, T. and Afroz, F. and Schöneich-Knipp, C. (1999). Liquid-liquid-liquid Microextration for SamplePreparation of Biological Fluids Prior to Capillary Electrophoresis, *Anal. Chem. 71,* 396R.

MacMurry, J. E. and Fany, R. C. (2008). *Organic Chemistry 5th Ed.* Pearson Education, Inc.

Martens, N. and Hall, E. A. H. (1994). Immobilization of Photosynthetic Cells Base on Film-forming Emulsion Polymers, *Anal. Chim. Acta, 292,* 49.

Peterson, E. M. Eisenthal, R., Danson, J. M., Spence, A. and. Daniel, M. R. (2004). A New Intrinsic Thermal Parameter for Enzymes Reveals True Temperature Optima, *J. Biol. Chem. 279,* 20717.

Peterson, E. M., Daniel, M. R., Danson, J. M. and Eisenthal, R. (2007). The Dependency of Enzyme Activity on Temperature: Determination and Validation of Parameters, *J. Biochem. 403,* 613.

Porath, J., Carlsson, J., Olsson, I. and Belfrage, G. (1975). Metal Cheleting Affinity Chromatography, *Nature, 258,* 598.

Radzicka, A. and Wolfenden, R. (1995). A Proficienty Enzyme, *Science, 267,* 90.

Answers to Exercises

Both numerical and short answers are given, some relatively easy essay types of answers are left for the students to attempt.

Chapter 1

1.11.1. (a) 1.238 (b) 2.1 (c) 3.01 (d) 2.00

1.11.2. (a) 3 (b) 4 (c) 4

1.11.3. (a) 14.7 (b) 74.5 (c) (d) 4.600

1.11.4. (b) 1.98 and 2.03, systematic; $\pm$ 0.01, and $\pm$ 0.02 random

   (c) 50.031, systematic; $\pm$ 0.009, random

   (d) random

   (e) Mass is systematically low because the empty funnel was not dried. In addition, there is a random error in every experiment, but we do not know exactly how much in one experiment.

1.11.6. (a) 2.1 $\pm$ 0.2 (or 2.1 $\pm$ 11%) (b) 0.151 $\pm$ 0.009 (or 0.151 $\pm$ 6%)

1.11.7. Systematic; low

1.11.8. 0.4507 $\pm$ 0.0005) M

1.11.9. 0.667 ($\pm$ 0.001) M

1.11.10. 31.998 $\pm$ 0.0006

1.11.11. 1.63 (($\pm$ 0.05) x 10^{-5}

1.11.12. (a) Percentage relative uncertainty in mass is 0.04 %, and percentage relative uncertainty in volume is $4._4$ %.

   (b) Density = 1.4 $\pm$ 0.2 g/ mL

     Whereby: (a) y = 2.129 (b) y = -1.927

Chapter 2

2.8.1. (a) 0.0010 moles; 0.002 moles (b) 20 mL

2.8.2. (a) 6.02 M (b) 6.85 molal (m) (c) 20.0%

2.8.3. (a) 469.0 mL (b) 466.2 mL

2.8.4. (a) 9.95 x 10^{-3} (b) 4.19 (c) 5.79

2.8.5. (a) 7.46 (b) 2.9 x 10^{-7} M

2.8.6. (a) 8.18 (b) 8.12 (c) 8.25

2.8.7. (a) pH = 13 (b) pH = 12 (c) pH = 12

2.8.9. cyanoacetic acid having pKa = 2.47 is the most suitable, because pKa is the closest to pH.

2.8.10. From given values of K_b, both K_a and pKa are obtained as follows: K_a = K_w/K_b and pK_a = -log K_a Ammonia having pKa = 9.28 is the most suitable, because pKa is the most closest to pH

Chapter 3

3.9.1. (a) The extent of dissociation of a weak electrolyte is decreased when a strong electrolyte containing an ion in common with the weak electrolyte is added to it. (b) If an external source of M^+ such as M^+Cl^- is added to a solution of M (aq), the equilibrium will always shift so that $[OH^-]$ decreases and increases, effectively suppressing the ionization of M.

3.9.2. CaF_2 will precipitate first

3.9.3. Because $Q > Ksp$, precipitation of Na_2SO_4 will occur

3.9.4. 7.08

3.9.5. (a) Increase (b) Decrease (c) No change (d) Decrease

3.9.6. (a) pH = 1.54 (b) pH = 7.00 (c) pH = 3.30

3.9.7. (a) $[OH^-]$ = 0.116 M, pH = 1.12 (b) $[OH^-]$ = 0.0.0125 M, pH = 12.10

3.9.8. (a) 1.143 x 10^{-3} (b) 0.0229 M

3.9.9. pH = 12.00

3.9.10. 1.1 x 10^{-10}

3.9.11. $Q_c > K_{spa}$ for CdS; CdS will precipitate; $Q_c < K_{spa}$ for ZnS; Zn^{2+} will remain in solution

3.9.12. pH = 4.04

Chapter 4

4.7.6 (a) A charged species crossing the phase boundary will create a potential difference between the phases that will tend to stop more ions of that charge crossing the boundary. The amount of charge that is necessary to achieve the impeding voltage is extremely small.

(b) The equilibrium condition is that the chemical potential of the species is the same in both phases.

4.7.7. Mixture contains with RCO_2^-, RNH_2, Na^+ and OH^-, all of them will pass through the cation-exchange column.

4.7.8. $NH_3 < (CH_3)_3N < CH_3NH_2 < (CH_3)_2NH$

4.7.9. The overall cell voltage is 0.701 V, this being positive the net reaction is spontaneous in the forward direction. Cd(s) is oxidized to Cd^{2+} and AgCl (s) is reduced to Ag (s). Electrons flow from the left electrode to the right electrode. If we could have found a negative voltage, the reaction would have been spontaneous in the reverse direction.

4.7.10. (b) 6.422 x 10^{18} e-/C (c) 9.649 x 10^4 C/mol

4.7.11. (a) 3.02 x 103 C (b) 0.840 A

4.7.12. (a) 0.012 (b) 1.7 x 10^{-10} (c) V_{org} = 17 mL

Chapter 5

5.8.1 (a) Absorbed impurities are taken into a substance and adsorbed impurities are found on the surface of a substance.

(b) Included impurities occupy lattice sites of the host crystal. Occluded impurities are trapped in a pocket inside the host.

5.8.2. (a) An ideal gravimetric precipitate should be insoluble, be easily filtered, possess a known, constant composition, and be stable to heat. (b) High supersaturation usually leads to formation of colloidal product with a large amount of impurities.

5.8.3. During the first precipitation, the concentration of impurities in the solution is high, giving a relatively high concentration of impurities in the precipitate. In the precipitation, the level of solution impurities is reduced, therefore giving a purer precipitate.

5.8.4. (a) Washing with electrolyte preserves the electric double layer and prevents perptization (b) The volatile HNO_3 evaporates during drying. $NaNO_3$ is nonvolatile and will lead to a high mass for the precipitate.

5.8.5. Supersaturation can be decreased by increasing temperature (for most solution), mixing well during addition of precipitant, and using dilute reagents.

5.8.6. 8.66 wt%

5.8.7. 0.023 M

5.8.8. 19.98 wt%

5.8.9. 75.40 wt%

5.8.10. (a) Increased mass of product could arise from co-precipitation of $CaSO_4$ with $BaSO_4$. (b) Longer digestion time removes $CaSO_4$ from the product.

5.8.11. 6.775 mol. Hg

5.8.12. A local excess of reactant concentration is required to initiate nucleation and form a small crystal. This is because the solubility of precipitate below 10^{-3} mm in diameter is greater than that of large particles. Therefore, to form the initial small particles, the final equilibrium concentrations must be exceeded.

5.8.13. 3.226 ppt Cl^-

Chapter 6

6.6.6. An indicator compound for a complexation titration must be a ligand with an effective formation constant for the coordination center smaller than that of the titration ligand (which is not too small), and its complexed and uncomplexed forms must be differently coloured.

6.6.7. The equivalence point for the Mg^{2+} titration is at a much more favorable position for Eriochrome Black T than it is for the Ca^{2+} titration. When both cations are present, the BET colour changes when both metals are titrated.

6.6.8. If the EDTA titrant contains the Mg^{2+} or if Mg^{2+} is added as MgY^{2-}, the end point will not be affected by Mg^{2+} ions.

6.6.9. Single coordination site ligands often form stepwise complexes, which greatly reduces the sharpness of the concentration change in the equivalence point region of a titration. The chelate can occupy all coordination sites of the coordination center with a single ligand, so that a 1:1 complex of generally high formation constant results.

6.6.10. An increase in the formation constant of the complex will cause an increase in the magnitude of the concentration changes in the equivalence point region. Conversely, a decrease in the analyte concentration will cause a decrease in the size of the concentration change in this region.

6.6.11. (a) 23 mL (b) 3.134%

6.6.12. 2.02 ethoxyl group / molecule

6.6.13. 61.1% of the sample

6.6.14. From Figure 6.7, we see that at low $p[H_3O^+]$, say 30, the formation constant for Fe^{3+} is about 10^{15}, while that for Fe^{2+} is about 10^4. EDTA would therefore complex selectively with Fe^{3+} at this $p[H_3O^+]$.

6.6.15. It is the formation constant of the metal ion's reaction with EDTA that determines the position of the left side of the K'_{eff} curve. The lower the formation constant, the more to the right the curve is shifted.

Chapter 7

7.7.1. The acid-soluble inorganic and organic matter can probably be dissolved (and oxidized) together by wet ashing with $HNO_3 + H_2SO_4$ in a Teflon-lined bomb in a microwave oven. The insoluble residue should be washed well with water and the washings combined with acid solution. After the residue has been dried, it can be fused with one of the fluxes, dissolved in dilute acid, and combined with the previous solution.

7.7.2 (a) most probable number of red balls = np_{red} = (1000) (0.06) = 60; most probable number of blue balls = np_{blue} = (1000) (0.44) = 440

(b) absolute; red = blue is 5.14

Relative standard deviation for red is 8.56% and blue is 1.17%

(c) 1.83×10^4

7.7.3. (a) 5.0 g (b) 7

7.7.8. The composition of a random heterogeneous material varies from place to place with no pattern or predictability to the variation, while a segregated heterogeneous material has relatively large regions of distinctly different compositions.

7.7.9. A random sample is selected by taking material at random from the lot. That is, the location from which the material is selected should follow no pattern. You could do this by dividing the lot into many imaginary regions and assigning a number to each. Then with the help of your calculator or computer generate random numbers and take a sample from each region whose number is selected by the computer. While a composite sample is selected deliberately by taking predetermined portions of the lot from selected regions, random sampling is appropriate for a random heterogeneous lot. Composite sampling is appropriate when the lot is segregated into regions of different composition.

Chapter 8

8.16. 1. 2.0×10^{-9} M; solution is a basic

8.16. 2. 2.3×10^{-7} M

8.16.3. (a) The titration of a weak acid B, produces the conjugate acid, BH^+, which is necessarily acidic.

(b) $V_e = 12.5$ mL, pH = 1.30, 1.35.

8.16.4. pH = 13.00 and pH = 12.68

8.16.5. pH = 8.18

8.16.6. $pKa = 3.72$

8.16.7. pH = 9.44

8.16.8. $V_e = 5.00$ mL, pH = 3.40 and 4.60

8.16.10. (a) Colourless to pink (b) systematically requires too much NaOH

8.16.11. $V_e = 47.79$ mL; pH = 8.74, 5.35.

8.16.12. 40 mL of 0.040 M NaOH is required to reach the equivalence point (a) 6.98 (b) 7.46 (c) 9.75

8.16.13. (a) An indicator compound for a complexation titration must be a ligand with an effective formation constant for the coordination center smaller than that of the titration ligand (but not too small), and its complexed and uncomplexed forms must be differently coloured.

(b) The indicator ligand may also hydrolyze, also the protonated forms and the colours of these forms must be considered when choosing an indicator for a specific application.

Chapter 9

9.6.2. 0. 2 mm

9.6.5. (a) There is no broadening by multiple flow paths in an open tubular column

(b) The longer the column, the better the resolution. It is always good to use a longer open tubular column than a packed column because the particles in the packed column resist flow and require high pressure for high flow rate.

9.6.6. (a) 4.3×10^4 (b) 3.6 μm

9.6.7. 6.3 cm diameter x 25 cm long, 17 mL/min.

9.6.8. (a) $w_{1/2} = 0.172$ min, N = 3.58×10^4 (b) 0.838 mm (c) w (measured) = 0.311 min; $w/w_{1/2}$ (measured) = 1.81

9.6.9. heptane : 7.4 x 10 x 10^4 plates, 0.40 mm plate height
       $C_6H_4F_2$: 8.6 x 10 x 10^4 plates, 0.35 mm plate height

9.6.10. (a) 1.11 cm diameter x 32.6 cm long (b) 0.069 mL (c) 0.256 mL/min

9.6.11. (a) lower (b) higher

9.6.12. (a) $u_{optimum}$ = 31.6 mL/min (b) $u_{optimum}$ = 44.7 mL/min $u_{optimum}$ increase

9.6.13. (b) At pH 3. The net flow of cations is heading to the right and the net flow of anions is heading to the left.

9.6.14. (a) 5.7 mL (b) 11.5 mL (c) adsorption occurs

Chapter 10

10.8.2. The enzyme, being a true catalyst, is not changed or consumed by the reaction. When it has catalyzed the conversion of one substrate molecule, it is free to facilitate the reaction of another. The cofactor or coenzyme is not necessarily a true catalyst and may be changed by the reaction. In this case, a second reaction (often enzyme-catalyzed) is needed to return the coenzyme to its original form so that it can precipitate in further reaction.

10.8.4. In addition, binding the substrate, the enzyme may need a cofactor to which it must be bound. If another reactant such as O_2 is required, it must also be bound to the enzyme. In some cases, the formation of the active site requires the formation of a multimer of piptide chains. In addition, some enzymes require an activator ion or effect or molecule.

10.8.5. Any buffer system with a p$K'a$ value within 0.5 pH units of the desired pH can be used for effective pH control. From Table 10.1, possible buffers would include MOPS, HEPES, and HEPPS. The one selected will have to be tested with the biochemical reaction to determine whether they inhibit the desired function. This is because the weak acid anion in the buffer may complex metal ions that are part of the enzyme and critical to its activity.

10.8.7. An epitope region of an antigen bonds to a paratope region of the antibody designed for it. The bond is a combination of van der Waals forces, hydrogen bonding, coulombic interactions, and hydrophobic interactions.

10.8.8. A uniform pore size in the membrane material used for filtration will create a sharp cutoff between the sizes of particles that will pass the filter and those that will not. Since this is a batch operation, the higher degree of discrimination can be a good idea. For gel permeation chromatography, a uniform pore size will produce a narrow range of sizes between the unretained and total access sizes. This will give better sensitivity for size dispersion in this range, but it provides a smaller range of sizes over which separation occurs.

10.8.9. If the ligand were not immobilized, it would not be possible to rinse away the sample components that had not become bound to it without rinsing the complex away as well.

10.8.10. The analyte is released by introduction of a component with a still higher affinity to the ligand than the analyte or by changing the solution pH or ionic strength so that the active site of the ligand is distorted and the formation constant for the complex reduced.

10.8.11. No, ligands in affinity chromatography are not limited to antigens or antibodies. Any biochemical species with the property of selective binding can suffice. Enzymes and other proteins are usually used.

10.8.12. The serum often produced is polyclonal because of the number of different haptenic determinants that have been introduced by one or more antigens.

10.8.13. At very high and very low pH's, enzymes will lose their tertiary and quaternary structure. The enzyme may not recover its active structure after a reasonable pH is established. Some of the amino acids in the peptide structure can change their acid form as a function of pH. This can cause a change in the shape and binding properties of the active site. This kind of deactivation is generally reversible.

10.8.14. $\Delta t = 2.2 \times 10^5$ s this many seconds is two and a half days.

10.8.15. The partial specific volume has the units of $cm^3 g^{-1}$. It is the reciprocal of the density, which has the units of $g\ cm^{-3}$.

Abbreviations

b = box   i = illustration
p = problem t = table
r = reference f = figure

Indexes

CURRICULUM VITAE OF THE AUTHOR

Name:	Wilson N. William
Date of birth:	**9th December 1964**
Place of birth:	**Muleba Tanzania**
Religion:	Christian
Nationality:	Tanzanian
Marital status:	Married with four children
Mailing address:	P.O.Box 370 SEKOMU Lushoto
Telephone:	0716457062 or 0688634297 or 0745880649

E-mail: ngambeki1964@gmail.com, william.wilson@sekomu.ac.tz

II: ACADEMIC QUALIFICATIONS

College/Institution	Major field of study	Year attended	Award
University of New Mexico U.S.A.	Organic, Organometallics Chemistry and Chemical Education	2004-2011	Ph.D.
University of Dar es Salaam	Chemistry	1993 to 1996	M.Sc. Thesis
University of Dar es Salaam	Chemistry, Physics and Ed.	1987 to 1991	B.Sc.Ed.(Hons)
Buhemba	National Service	1986 to 1987	National Service Certificate
Mkwawa T.T.C.	Chemistry, Physics and Ed.	1985 to 1986	Diploma in Ed.
Mkwawa High School	nistry, Physics and Ed.	1983 to 1985	Advanced Certificate of Secondary School

Kashozi Secondary School	ndary Education (Science)	1979 to 1982	Certificate of Secondary School
Mugoma Primary School	Primary School Education	1972 to 1978	Primary School Leaving Certificate

III: PROFESSIONAL/ WORK EXPERIENCE

1. **From 2014 to-date: Appointed as senior lecturer of Chemistry**
2. From 2012 to-2014: Appointed as lecturer
3. From 2004 to-2011: Working as Research assistant and Tutorial assistant

IV: ADMINISTRATIVE/LEADERSHIP POSITIONS HELD

1. From 2016 to-date: Acting As Deputy Vice Chancellor for Academics, Research and-Consultancy (DVC-ARC)
2. From 2016 to-date: As DVC-ARC I am a coordinator of the project with a title: Capacity Building on the Rights of Education for Children with Living with Disabilities around Usambara Mountains with the Sponsorship of the Finish Evangelical Mission (FELM).
3. From 2015 to-2016: Appointed as SEKOMU coordinator of Integrated Teacher Training (ITT)
4. From 2014 to date: Appointed as staff representative to the Council
5. From 2013 to-date: Appointed as a member of the SEKOMU Senate
6. From 2013 to-date: Appointed as a member of the Senate Committee for Academic and Curriculum Affair (SCACA)
7. From 2012 to-2016: Appointed as Dean Faculty of Science
8. From 1998 to 2004: Head of the Department of Chemistry ST. Anthony, Sec. School
9. From 1993 to 1997: Head of the Department of Chemistry and Academic Master at Tambaza, Sec. School
10. From 1991 to 1993: Head of the Department of Chemistry at Azania, Sec. School
11. From 1986 to 1986: Tutor at Musoma Teachers' College

V: MEMBERSHIP TO PROFESSIONAL SOCIETIES

1. American Chemical Society (ACS)   Chemistry since 2008 to date.

2. Editorial Board Member of International Journal of Advances in Chemistry (IJAC) as well as PCM member of Chemistry Conference of AIRCC worldwide since 2016 to date.

VI: PUBLICATIONS

1. ***William, Wilson N.*** *Synthesis and Crystal Structure Characterization of Ethylnitriletrispentafluorophenylborane , Acta Crystallographica Section E.* **In preparation 2016.**

2. ***William, Wilson N.*** *Synthesis and Crystal Structure Characterization of 2,4,6-Tri-tert-butylbenzenaminium iodide, Acta Crystallographica Section E.* **In preparation 2016.**

3. **William, Wilson N.** *The Multi-electrolyte Aquatic Environment of Africa East African Rift Valley Lakes: Influnce on Chemical Equilibria and Buffer Capacities, Tanz. J. Sc.* **Submitted, 2016**

4. **William, Wilson N.** The Fluoride and Trace Metals Contents in Lift Valley Lake Eyasi Salt a Possible Threat to Human Health , *Huria Journal,* ***Accepted and Submitted May, 2016.***

5. ***William, Wilson N. and Joseph K. Ho, 2015.*** *The Effectiveness of Using Calibrated Peer Review (CPR) Against Non-CPR (Traditional) Means in Submitting Chemistry Laboratory Reports, Huria Journal,* **19**, *131-145.*

6. ***William, Wilson N.* 2014.** *Synthesis and Crystal Structure of Diguandinium diaquapentakis (nitrato)neodymiate(III), J. Struct. Chem.* **55**, *1178-1181.*

7. *Agness Mrutu,* ***William, Wilson N.*** *and Richard A. Kemp,* **2012.** *Synthesis and Characterization of Molybdenum and Tungsten Complexes Containing Tris(diphenylphosphino)methane (tdppm). Inorg. Chem. Commun.* **18**, *110-112.*

VII: BOOKS

1. **William, Wilson N.** *Integrated Challenging Conceptual Chemistry Laboratory*

   Experiments and Techniques, **in Preparation, 2016.**

2. ***William, Wilson N.*** I. *The Synthesis of 6π-Electron Analogous Phosphine Types Ligands, Published with* **ISBN: 978-3-659-58251-6** Lambert Academic Publishing, Germany, **2014**.

3. **William, Wilson N**. M.Sc. Thesis A Study of the Chemical Constitution and Hydrochemical Equilibria in a Concentrated Salt Lake: Lake Eyasi, University of Dar es Salaam, **1996**.

VIII: MANUALS
1. **William, Wilson, N.** Organic Chemistry Practical III, **2012.**

IV: CURRICULA PREPARED
1. **William, Wilson, N., Emmanuelina Kateme, and Wilson Namasaka,** Bachelor of Science with Education (BSc. Ed.) 2013.
2. **William, Wilson, N., and Linus Munishi,** Master of Science in Nature Conservation and Management, (MSc.NCM) **2014.**
3. **William, Wilson N., and Gaudence, Mohode** Bachelor of Science Computer Science (BSc. CS.) **2014.**
4. **William, Wilson N., and Gaudence, Mohode** Diploma in Computer Science (Dip. CS.) **2014.**

X: POSTERS/PRESENTATIONS
1. **William, Wilson N.;** Ho, Joseph K. **Feasibility of Using Calibrated Peer Review™ (CPR) in the General Chemistry Laboratory.** Success in the Classroom: Sharing Practices that work. The sixth Annual UNM Community Conference for Teachers by Teachers University of New Mexico USA **2011.**
2. **William, Wilson N.;** Kemp, Richard. **Synthesis and coordination chemistry of new 6π-electron ligands** . A poster presenter at 237th American Chemical Society (ACS) National Meeting, Salt Lake City, UT, United States, March 22-26, **2009,** INOR-566. Language: English, Database: CAPLUS
3. **William, Wilson N.;** Mosha, D. **S Influence of concentrated aqueous electrolyte environment on Dissociation constants (pKa,s) for phosphate and the evaluation of buffer capacities in such media.** Kampala **1995** proceedings are currently on World wide web (internet) for global scientific community.

4. **William, Wilson N.;** Mosha, D. S **The evaluation of pH, pKa,s and buffer capacities in the natural concentrated mult-electrolyte environments of East Africa Salt lakes.** A paper presented at international workshop Nairobi **1996.**

XI: SUPERVISION OF SPECIAL PROJECTS AND THESIS
BSc. Projects
1. Ecological Impacts of Water Hyacinth on Lake Victoria Ecosystem Kuruthum, H., **Completed in 2013.**

2. Identification of Medicinal Plants at MagambaVillage (Michael, A., **Completed in 2013**.

3. The Problem Facing Eco-waste Management in Lushoto Town Thobias, J., **Completed in 2013**.

4. "Assessment of water Quality" A case Study of Lushoto District Mhagama, B., **Completed in 2013**.

XII: INTERNATIONAL CONFERENCES ATENDED

1. 21-8-2015 – Iternationalization Through Linnaeus-Palme Foundation Annual Conference in StockholmHeld at Swedish International Development Cooperation Agency (SIDA) Headquarter.

2. 28-8-2013 – LEPUS Closing Conference (SUA, KUL and SEKOMU).

3. 7-11-2013 – International Scientific Conference on Mental Health (MEHATA)- TANZANIA.

4. 15-11-2013 – Joint Symposium on the social impact on the United Nations Convention on Right of Person with Disabilities (together with EFH Bochum- Germany).

5. 22-3-2009 - 237th American Chemical Society (ACS) National Meeting, Salt Lake City, UT, United States of America.

XIII: AWARDS

1. Given a certificate of attendance for having successfully attended and actively participated in **Higher Education Teaching Methodology** Course conducted on October, SEKOMU, (**2012**).

2. Being the best Tutorial Assistant (TA) for the year **2009** and awarded a scholarship of **$ 15,000.00** by the Faculty of Art and Social Sciences University New Mexico, Albuquerque USA. (**2009**).

3. Tanzania Chemistry Panel member, elected by the Director of Tanzania Institute of Education from the year **2000 – 2003**, to deal with Chemistry curricula for O-level and High Schools.

4. Given certificate of appreciation for excelling in Chemistry following my form six students being number nine (9) out of seventy four (74)

secondary schools in Tanzania whereby 100% of the students passed in the year (**2003)**.

5. Given certificate of appreciation for excelling in Chemistry following my form six students being number nine (9) out of seventy four (74) secondary schools in Tanzania whereby 100% of the students passed in the year (**2002**).

6. Given certificate of appreciation for excelling in Chemistry following my form six students being number nine (9) out of seventy (70) secondary schools in Tanzania whereby 100% of the students passed in the year (**2001**).

7. Given certificate of appreciation for excelling in Chemistry following my form six students being number eight (8) out of fifty four (54) secondary schools in Tanzania whereby 100% of the students passed in the year (**2000**).

8. Given certificate of appreciation for excelling in Chemistry following my form six students being number fourteen (14) out of fifty four (54) secondary schools in Tanzania whereby 95% of the students passed in the year (**1999**).

9. Given certificates of being the best student in Basic Mathematics, Chemistry, Geography, and Book keeping during form four (IV) graduation at Kashozi Secondary. School in (**1982**).

www.ingramcontent.com/pod-product-compliance
Lightning Source LLC
Chambersburg PA
CBHW080936120726
48003CB00011B/3181